Y0-DJN-246

Dr Jerry LeMieux

Introduction by Maj General (Ret) James O. Poss

Cover Photo is of Stephen Raleigh testing one of his homebuilt UAVs

ISBN **978-0-578-12880-1**

Printed in USA by Unmanned Vehicle University Press
www.uvupress.com

Contents

Preface

From horse carriage to car, from standalone PC to internet, drones will revolutionize the aviation industry. There are already successful commercial drone businesses all around the world. The professional community calls them Unmanned Aircraft Systems (UAS) or Unmanned Aircraft Vehicles (UAV). The military calls them remotely piloted vehicles (RPVs). For the purpose of this book, we will call them the more popular term "drones." Drones are currently in an emerging commercialization phase and those that start businesses now could benefit by being the first. Interest is growing in civil uses, including commercial photography, aerial mapping, crop monitoring, advertising, communications and broadcasting. Unmanned aircraft systems may increase efficiency, save money, enhance safety, and even save lives[1].

There are numerous established commercial drone businesses around the world. In Canada the company Accuas (accuas.com) has been doing professional ground based site surveys that normally take two weeks to complete. Using a drone, the same site survey can be completed in two days. In the UK, CYBERHAWK (thecyberhawk.com) is the first company in the world to use drones to inspect offshore oil rigs reducing costs and keeping the rig online. And finally, Dominos is testing pizza delivery in the UK.

Drones have gotten a bad rap due to their current uses in the military and police work. I am going to make one point about the military and then move on to commercial applications. I watched the movie, Born on the Fourth of July last night and thought about the nearly 50,000 combat fatalities that occurred during the Vietnam War. I started to think about comparing past casualty numbers with the current war on terror. From October 7, 2001 through May 29, 2012 the US had 5,078 US combat fatalities[2] from the Global War on Terror.

Table 1 lists the number of combat fatalities from WWI to Iraqi Freedom[3]. It is difficult to make a comparison for many rea-sons but you can observe the length of time for the Vietnam War is similar to the time we have been involved in the Global War on Terror.

If you compare 5,078 casualties to Vietnam casualties you can conclude that for approximately the same time period, there were ten times lower combat fatalities. One significant reason for this dramatic reduction in casualties is drones. On Feb 21, 2013 Senator Lindsey Graham, who is a member of the Senate Select Committee on Intelligence, revealed to a North Carolina Rotary Club that drones have reduced casualties. He said "First, there is no pilot on board; the drone takes all the risk.

War	Years	Combat Casualties
World War 1	1017-1918	53,402
World War 2	1941-1946	291,557
Korean War	1950-1953	33,741
Vietnam War	1961-1973	47,415

Table 1 US Combat Fatalities from WWI to Operation Iraqi Freedom

American casualties have always been the obstacle to waging war, which is of course what you have to do to save lives. However, the American people will allow only so many combat deaths for the sake of saving American lives. Drones eliminate American casualties as an obstacle to waging war."That is the last you will hear about military applications.

We are now in an era where drones are moving from military to commercial applications. This book focuses on the good things that drones can do for society. I have analyzed many industries for potential commercial opportunities. The three criteria for my analysis are: does a drone save time; does a drone save money over a manned aircraft or ground application; and/or does a drone reduce risk to human life. The result is Appendix 1 where I list 200 commercial applications for drones. The list contains what you can do with a drone. I narrowed this list down to 30 business that have the highest potential to start a business and make a profit.

The How to Start a Drone Business Chapter describes the investments you will have to make and training you will require to prepare for the hundreds of commercial opportunities that are coming available.

The obstacle to commercial drone operations in the USA is lack of regulations. Another obstacle that is often sighted is the potential for collision between a drone and an aircraft. I believe that this is a non-issue for small drones that fly at very low altitudes. In Canada, the UK, and Australia, regulation for commercial operation of drones is already available. In many countries, you don't need a pilot's license or airworthiness certificate if the drone is below 20 kg and flies at low altitude. Federal Aviation Regulation (FAR) 91.117, "minimum safe altitudes" states that pilots are not allowed to fly below 1000 feet in congested areas and 500 feet in other than congested areas. If you think about it, a small drone will need to get very close to what it is using the sensor to image. There will easily be hundreds of feet of separation between a low flying aircraft and a small drone flying at low altitude. So there is inherent air traffic separation due to the 91.117 regulation and the fact that you will need to fly at very low altitudes to get good imaging with small sensors. FAA Advisory Circular (AC) 91-57 states that you can fly a model aircraft as long as you stay below 400 feet, remain three miles away from an airport and avoid populated areas.

Low flying model aircraft are so safe, that if you join the Academy of Model Aeronautics (AMA) you get $2.5 million of liability insurance. There are 150,000 AMA members that have flown model aircraft for over 30 years without problems. Perhaps this regulation can be adapted by the FAA for small low altitude drones. Anyone can fly model aircraft for recreation or pleasure but the FAA draws the line when you charge for your services. You can fly your small drone around and take pictures, you just cannot sell them to anyone. Several businesses have violated this policy that is not in writing anywhere and have either been served a cease and desist order or fined. I belive if you want to start a commercial drone business there is no issue to practice without charging. There is a whole workflow you will need to setup to create a successful business including the purchase of drone and a sensor, figuring out how to to data processing and how to format your data into a final report you can charge for. The time to prepare your business is now, not after the FAA passes the regulations. If you wait until then, you will fall behind. First to market is a definite advantage.

Figure 1: Insitu Scan Eagle®: Endurance 24 hours, Wingspan 10 ft., Length 5ft, 20,000 ft., 48 knots, 44 lb., Catapult launch, Sky hook recovery, Optical/IR Sensors. Photo Courtesy of the U.S. Marines

At the time of the writing of this book, the Insitu Scan Eagle, shown in Figure 1, was the first in the USA to be approved for commercial operations[4]. In 2012 Congress passed legislation for the Bering Strait, portions of the Bering Sea and Arctic Ocean to serve as test areas for commercial drone use. The Scan Eagle is the first drone in the USA to be operated commercially.

The commercial drone business is shaking out into several major sectors. There are approximately 1000 companies in the USA manufacturing the aircraft, sensors or parts. I belive the drone industry is fragmented by lack of integrators. A significant opportunity exists in the systems engineering field or integration and test. In this book, I have chosen 30 businesses that I believe would be most profitable and I focus on expanding a description and implementation of each of these Businesses.

It may surprise you, but drones are already being used for certain government non-military applications in the USA. The USDA and Bureau of Land Management use drones to protect wildlife. NASA uses drones to monitor hurricanes and many drones have been used in search and rescue operations (and more recently to help investigate a murder). On the commercial side, Airware, a Silicon Valley startup just raised $11 million. They build autopilots for commercial drones. DroneDeploy is another startup that is developing the software that companies will use to control and manager a fleet of commercial drones. They want their software to be compatible with as many drones as possible. Those that start and establish businesses now will be the future leaders in this certain to revolutionize aviation industry.

The FAA announced in August 2013 that a draft small UAS rule would be out by the end of 2013 for public comment. Passing of this regulation will open the floodgates and allow numerous drone businesses to startup and generate thousands of jobs. It could become a trillion dollar industry and I believe commercial revenues will well exceed present and future military spending. Military spending is in decline and commercial spending is rising.

Appendices 1-6 are on FAA regulations and drone certification. Appendix 7 lists 200 commercial applications for drones. Appendix 8 gives information on the Unmanned Aircraft Professional Association (UAPA) which provides you with insurance, legal representation and business startup advice. Appendix 9 provides contact information for small drone sensor companies and Appendix 10 for small drone manufacturing companies. Appendix 11 provides a list of small drone component companies and Appendix 12 provides contact information for small drone services companies. Appendix 13 is on insurance providers and Appendix 14 describes the organization that published this book: Unmanned Vehicle University Press.

This book will be periodically updated with more commercial business applications. If you would like to contribute a Chapter or have your commercial drone product, service or business included in the next edition of this book send an email to submit@uvupress.com

Dr Jerry LeMieux
President & Founder Unmanned Vehicle University

Introduction

By Maj General James O. Posss

Congress expects the Federal Aviation Administration to allow unmanned aerial systems (UAS) to fly in our national airspace in 2015. It's the year our skies – and probably our lives – will change forever. It's also the year the greatest boom in the aviation market of this new century will begin. Reasonable men can differ over the size of this new aviation market. Some say it will be worth tens of billions, others think it might be trillions. Both sides agree it will be big.

The irony of this new market is that unmanned aerial systems aren't new. "Drones" have been with us since the beginning of aviation. Both sides had flying bombs during World War I. During World War II, the US Army Air Force lost a Kennedy brother and wounded Dwight Eisenhower's son during an unmanned B-17 bomber mission. "Buffalo Hunter" drones were often the only reconnaissance aircraft that could survive North Vietnamese air defenses. Israeli drones were the unmanned heroes of Israel's first Lebanese war in the 80's.

Unmanned systems may have been with us since the dawn of aviation, but it took another revolution to bring us to the brink of flying UAS safely in our national airspace. Modern unmanned systems that can fly safely alongside airliners need the same computer and communication revolutions that transformed the cell phone from a suitcase sized device that could only make phone calls to the palm sized smart phone that could do everything from making calls to making movies to guiding you to the perfect restaurant in an unfamiliar city. Without modern computer and communications horsepower, all previous unmanned aerial systems were either essentially larger versions of RC model aircraft that required a human pilot to be within eyesight to the control the aircraft or had simple autopilots that could only fly simple courses for relatively short periods of time. All this changed when the first modern UAS – the MQ-1 Predator – harnessed the computer and communication revolutions when it took to the skies over Bosnia in 1997

The Predator was a revolutionary UAS because it harnessed the same technology that made smart phones so brilliant. A Predator's pilot no longer had to be within sight of his air vehicle to control it. The Predator used communications technology similar to smart phones to access communication satellites to allow the Predator to fly thousands of miles away from its ground data link.

The same internet technology that brings email to your iPhone allowed a Predator to access ground data links to fly Predator from anywhere on earth. Unlike previous unmanned systems, the Predator wouldn't crash if it lost its control link. The same GPS technology that powers Google maps on your Android phone let the Predator find its own way home.

The US military quickly seized on this unmanned revolution to make UAS of all shapes and sizes. Within four years of the Predator's first combat mission, the airliner-sized Global Hawk flew over Afghanistan for the US Air Force. Two years after this flight the US Army miniaturized UAS technology into the hand launched Raven UAS. Just a few months ago, the US Navy landed its UCLASS UAS on an aircraft carrier. The Air Force now trains more UAS pilots than it trains fighter and bombers pilots combined. Many analysts even believe that the Joint Strike Fighter will be the last American manned fighter aircraft.

But if UAS aren't new and the Department of Defense has so enthusiastically adopted unmanned systems, why can't we fly commercial unmanned systems in our national airspace system? Much of the reason has to do with the rightly conservative attitude that the Federal Aviation Administration has towards any potential increased threat to aviation safety. UAS advocates may point to the fact the Air Force Predator now has a lower accident rate than the F-16, but the FAA will point out that UAS bring safety vulnerabilities that manned aircraft don't have. Manned aircraft don't need datalinks to operate. A UAS does. Manned aircraft don't have to deal with the complications involved with remoting a pilot's cockpit from his aircraft. A UAS does. Above all, pilots in manned aircraft can always just look out their window to avoid oncoming air traffic. A UAS can't.

But, assisted by Congress, the FAA does have a plan. By then end of 2013, the FAA will have six new test sites and a new research center of excellence to help them research how to safely fly UAS in our national airspace. My university, Mississippi State University, is a proud member of the Mississippi/Louisiana Test site team and is the lead university in a consortium of twelve other leading UAS research universities bidding for the research center of excellence.

We're uniquely suited to research how to certify UAS to fly in national airspace, how to give them the ability to "sense and avoid" hazards, how to make their data links larger, more secure and more reliable, how to present information to a remoted pilot without overwhelming them with data, how to assess environmental impacts and how to do it all without violating our right to privacy. These are all legitimate problems to solve before we should all be comfortable with UAS flying alongside airliners.

UAS are particularly important to us on the Gulf Coast of Mississippi and Louisianna because we know that being unable to fly UAS in our airspace has probably cost us lives and property. We were unable to fly UAS after the worst storm in US history hit us head on. Three years later we were unable to track our country's largest oil spill with UAS. And, when you consider that 43% of the landmass of the United States drains through our states, every gallon of fertilizer saved by UAS-aided precision agriculture means a smaller dead zone in our Gulf off the mouth of the Mississippi.

These are only three of the 200 commercial applications Dr. LeMieux envisions (Appendix 7) for civil UAS after we can fly them in our national airspace. I don't know if he has that number right. I do know that after I bought my first cell phone, I thought that it would only do three things – make calls, send text and take voice mails. I do know that I was off by an order of magnitude because I couldn't even envision a phone with GPS, a megapixel still camera (with a really cool panorama feature), a video camera, embedded accelerometers, a key board, microphone and a 4 megabit internet connection. I do know I have 54 apps on my iPhone, with another 899,946 to choose from on the App Store. I do know that the sky will literally be the limit when we start launching the equivalent of winged smart phones into our national airspace. And as a Gulf Coast native, I do know that lives will be saved in addition to fortunes made in this upcoming unmanned boom. Like I said in the beginning- the jury is still out if this will be worth billions with a "B" or a trillion with a "T"

Major General (Ret) James O. Poss

The FAA

Before you start operating drones in US National Airspace (NAS) you must be aware of the Federal Aviation Regulations (FARs). I have found that very few really understand the rules so I decided to include this Chapter. Appendix 1 contains the model aircraft Advisory Circular 91-57. This is not a regulation, it is an Advisory Circular (AC) and has been around since 1980. There are differences between a model aircraft and a drone, primarily the data link and a sensor. Many are interpreting this as a regulation for drone operations. It turns out that this is not a correct assumption and you could get in trouble if you operate under this assumption. The FAA could send you a cease and desist order or fine you or even take you to court. The primary reference for UAS operations is National Policy Document N8900.227 dated July 13, 2013. Part 8(b)(1)(b) states: "This notice and the processes prescribed do not apply to hobbyists and amateur model aircraft users when operating unmanned systems for sport and recreation. Those individuals should seek policy under the current edition of advisory circular (AC) 91-57, Model Aircraft Operating Standards.

FAA conops for drones fall into two categories. The first category is small drone operations where the weight of the drone is below 55 pounds. It is possible that you may not need a pilots license. Canada, the UK and Austrailia only require a permit to operate. You will most likely need to take a general knowledge exam and demonstrate that you can safely operate a drone. The FAA has announced they will have rules for this class of drone published for comment by the end of 2013 and hopefuly finalized sometime in 2014. The second categroy for operations will be to fly in the same airspace as manned aircraft. You will need a pilots license to operate and your rating will depend on several factors. For exapmple if you operate at night or in the weather, you will need an instrument rating. The rest of this section focuses on the second category.

FAA AC 91-57 is not to be used as a basis of approval for operation of any other aircraft, including by Federal, State, and local governments, commercial entities, or law enforcement." Drone operations in the USA are covered in the FAA National Policy document N8900.227. It has such things as pilot and observer qualifications and training requirements. Always check the FAA website to get clarification on any misunderstandings you may have. You can also call the Manager of the FAA's UAS Integration Office at 202-385-4835.

According to the FAA UAS Fact Sheet posted on the FAA website there are two ways you can legally operate a drone in the NAS. The first is to obtain a Certificate of Authorization (COA) and the second is to obtain an experimental certificate. You also need to distinguish between public and civil operations. Public operations basically refer to City, State or Federally owned drones and civil is everything else. It used to be that only government agencies (public) and not for profit universities could obtain a COA. As of July 31, 2013, Section 9 of N8900.227 now says that a COA can be obtained for civil and commercial operations.

The rules for applying for an experimental certificate are contained in CFR 91.125 (Appendix 2) and operating limits are contained in FAR 91.319 (Appendix 3). Note that 91.319 says you cannot carry persons or property for compensation or hire. According to 91.319 you can charge for your services as long as you don't carry persons or property. If you carry a payload, that is not persons or property and you could legally be compensated for your services.

Everyone is waiting for the key date of Sept 2015 for passage of regulations for drones as the milestone for when they can charge for services. As of August 2013, this is not a correct assumption. In that month the FAA approved a restricted type certificate for the AeroVironment PUMA and Insitu Scan Eagle drones. The FAA also approved ConocoPhillips for commercial aerial survey operations in Alaska. You can read all about restricted type certification in FAA Order 8110.56A.

You will need to fill out FAA Form 8110-12 to apply for the restricted type certificate. Once you have a type certificate you can operate in accordance with CFR 21.25 (Appendix 4). Note Part 2 which opens the door for you to start a business and conduct the special operations that are listed. The restricted type certificate for the Scan Eagle is contained in Appendix 5 and the certificate for the PUMA in included in Appendix 6. These are examples of commercial drone certification.

I recommend that you take a look at one other regulation called FAA National Policy Document 8130.34C, Airworthiness Certification of Unmanned Aircraft Systems and Optionally Piloted Aircraft. In order to obtain a restricted type rating you must obtain a special airworthiness certificate. You can apply for this by filling out form 8130-7. Finally you will also need to obtain a Registration Certificate and Aircraft Flight manual.

Most drones don't have flight manuals so if you decide to become a manufacturer and want certification you should get an operations manual together. By the way, I did not make this up, everything in this Chapter came directly from an FAA representative.

In summary, to operate a drone legally in the NAS, you will need to get a special airworthiness certificate, restricted type certificate, registration certificate, a commercial COA and a flight manual to be legal to fly in the NAS. Don't forget to scan N8900.227 to make sure you are qualified as a pilot. Or you can save yourself all the hassle and buy a Scan Eagle or PUMA and apply for a commercial COA. If you fly a small drone below 55 pounds, you may not need a pilots license but will need to take an exam and demonstrate flying proficiency.

The Trends

In the past, large vehicles were required to lift heavy sensors into action. Today, drones are getting smaller and more affordable. The sensors are so small they fit into the palm of your hand. Small EO/IR sensors include the UAV Vision CM 100 (700 g), IAI Micro Pop (1.5 lb.) and the FLIR Tau (130 g). Small hyperspectral sensors include the Headwall Photonics Micro Hyperspec (1.8 lb.) and the Rikola (300g), the world's smallest). Vegetation stress cameras used for agriculture applications operate in the optical and NIR bands and can be bought off the shelf as a small digital camera. Tetracam makes the ADC lite (520 nm – 920 nm) and Cannon XNite ELPH320NVDI.

Figure 2 UAV Vision CM 100 EO/IR (700 g)

A major reason for the coming explosion in drone opportunities is the reduction in distance between the sensor and the target. Large drone systems need to standoff over hostile environments. Small commercial drones may operate as low as 50 feet and do not require large size, weight and cooling. As a result the small drones are catching up with the large drones in terms of performance. The small Penguin UAV manufactured by UAV Factory set a world record of 54 hours and 27 minutes in 2012. This UAV weighs in at 21.5 kg and has a 10 foot wingspan. The endurance is greater than any US military UAV. With the reduction in size, comes a reduction in price. Affordability is the driving force for the upcoming explosion in business creation and jobs. Many remote sensing projects have been done with satellites in the past. Now we are entering an age where sensing will be accomplished with drones at a more precise centimeter size resolution, less susceptibility to weather interference and faster turnaround times.

Figure 3 Penguin UAV. Courtesy of UAV Factory

What's going on with drones is similar to what happened to the hamburger. We cut up the hamburger into small pieces and gave it a name, "slider". In the same way we cut up an airplane by removing the dual instruments, dual flight controls and human environmental controls and gave it a name, "drone" One of the major advantages is very long endurance times and other advantages are significantly lower costs than manned aircraft and robust design features such as a drone that you can láunch from your hand after engine start. You simply throw it into the air. The main components are the air vehicle, a payload (sensor or cargo), data link and ground control station.

The brain of the drone is the autopilot. This is a hardware component that requires software to control various functions. There are many sensors that are tied into the autopilot such as the altimeter, accelerometers and gyros. Payloads come in a variety of shapes and sizes. In most cases the payload is a sensor such as a camera, infrared, multi-spectral, hyperspectral, ultrasound and others. A payload can also be some type of cargo to be delivered to a location. There are actually 5 different data links. The first is the uplink which controls the air vehicle. The second is the uplink that controls the payload. The third is a downlink that transmits payload data. The fourth is a downlink that transmits vehicle status and the final link is GPS. Finally, there is a ground control station. For a small drone, this can either be a manual control set or a laptop with external controls for flight or a point and click feature to designate points for the flight path.

When you start a drone business the first step is to choose an application. Then you need to optimize the sensor that fits the application. Put the air vehicle and sensor together an go out and test your theory. Gain experience about what you are seeing from the sensor and the processed data. Then accumulate lots of data into a report you can give to your customer so a decision can be made. This is the workflow process to start a drone business.

How to Start a Drone Business

Gene Payson, President Troy Built Models

There are numerous business opportunities for UAVs. A list of 200 potential commercial applications is shown in Appendix 7. The first step in starting a UAV business is to choose one of the commercial applications that you are interested in. Next, perform a market analysis for how the application is currently accomplished with manned, satellite or airborne platforms. Next, do a business case analysis to determine if there is potential to make a profit. You need to understand the science of remote sensing and decide which sensor will optimize data collection for your application. And finally you need to be trained as a UAV pilot

The Market

Current estimates for the UAV market are at $80 billion for the next 10 years worldwide. This amount translates to tens of thousands of new businesses and new jobs. Most estimates are for military applications and do not include commercial systems. The Teal Group estimates that 95% of commercial UAVs will be small. In General Poss's opinion, once commercial UAV operations start in the USA, it will be hundreds of billions with a "B" to a trillion with a "T" industry. So there is outstanding growth potential and an opportunity for you to get in on the ground floor in this exciting industry.

The Regulations

Many countries in the world already have regulations for commercial UAV operations including Canada, UK and Australia. In the USA the regulations are coming soon. In the USA, a small UAV is defined as less than 55 pounds.

In other countries it is less than 20 kg. Congress requires that the FAA develop regulations for small UAVs by August 2014 and for all UAVs by Sept 2015. The time to start preparing is now, not when the regulations are in place.

Small UAV Business

A small UAV business owner could have a gross income of about $80,000 to $120,000 per year. There are two approaches to operations. One is to become an expert at all things including being a UAV pilot and also analyzing the data and creating a report for decision making. A second approach is to team up with a specialist in your application area. If you decide to choose an agriculture application, you could fly the drone, capture the data and hand it off to the specialist for analysis and reporting. After obtaining the results you can publish a report and deliver it to your customer for a fee.

Professional Photographers of America (PPA)

The PPA was started back in 1869 and educates photographers on business and technical skills, provides a network of individuals in the industry and protects photographer's rights. PPA is a great resource for training, certifications, competitions, contracts, forms, copyright laws, technical tips and tricks, legal representation, marketing, pricing and networking.

Aerial Photography

There are many opportunities to make money with a drone using a commercial off the shelf camera. Some commercial applications that already use aerial photography are: hotels, skyscrapers amusements, boats for sale, condos, marinas, golf courses, sports stadiums, bridges, malls and skylines. Sample charges are $200 per set of photos or $1500 per day. There is a company called stockpix.com that charges by the print. Your customer could be someone that needs a stock photo for a magazine or cover for a book. You could take some of your photographs and display them on your website for purchase. Average charges are $35 to $230 per print. A small drone from DJI Innovations only costs $699 and can be found on the Troy Built Models website.

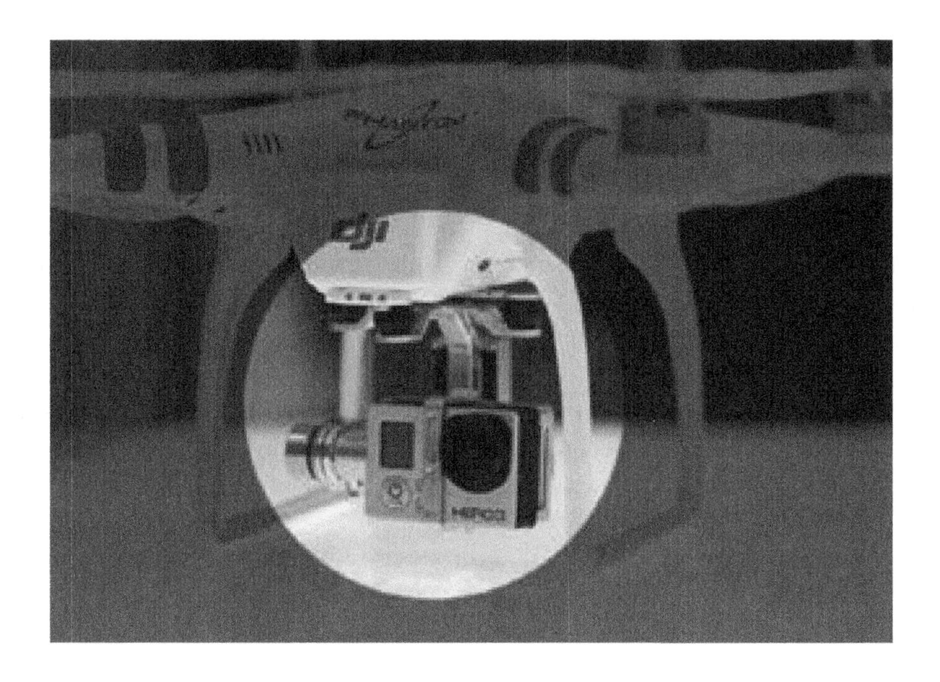

Figure 4 DJI Phantom. Courtesy of Troy Built Models

Take a look at other companies that are doing this business around the world to see how they do their marketing and get their customers. A company in Australia called uavsystems.com does aerial photography with a variety of platforms. Their platform endurance times vary between 30 minutes and one hour. A large unmanned helicopter can actually fly for 2 – 5 hours. Another company called Pixy UAV uav-aerial-photo.com uses a parasail drone with a gas engine. They will actually train you on their drone for around $500 Euros per day. Another company called Big Mountain Robotics www.dronemapper.com provides mapping and survey services. Film FX and MI6 Films www.mi6.com use drones to make videos for the filming industry.

Another company in Florida called Sunshine Sky Pics www.sunshineskypics.com has customers in commercial and residential real estate, construction and property developers, landscape architects, roofing, golf courses, marinas, campgrounds, natural preserves or parks, insurance claims, special events, advertising, marketing and websites, mapping and surveying companies.

For residential real estate they charge $109 for 10 photos, for commercial real estate, $189 for 10 photos and for high altitude shots using manned aircraft, $299 for 10 photos (renting a Cessna 172 may cost around $150 per hour). There is an individual in Florida that takes pictures of the damage done to roofs for insurance companies. The small company is busy enough to make this a full time endeavor. So if you focus on an area and become an expert, you could get all the business you would ever need.

The Department of Natural Resources will pay you to do surveys of invasive plants. NOAA has contracted civilians to fly a drone from a ship in the Aleutian Islands to count the sea lions.

The primary advantage of a drone is cost. Manned aircraft charge between $400 - $1500 per job. Other advantages are ease of setup, time efficiency and flight between 0 – 500 ft. A vertical takeoff and landing (VTOL) drone does not require a runway. While hovering. multiple shots can be taken from the same position and can capture a unique perspective that other methods can't.

Join Organizations

The Association of Private Photogrammetry, Mapping and Geospatial Firms (MAPPS), www.mapps.org is the only association of firms in the surveying, spatial data and geographic information systems field in the USA. MAPPS is engaged in surveying photogrammetry, satellite and airborne remote sensing, aerial photography, hydrology, aerial and satellite image processing, GPS and GIS data collection and processing services. The American Society for Photogrammetry and Remote Sensing (ASPRS) was founded in 1934. They promote application of active and passive sensors, the disciplines of photogrammetry, remote sensing and geographic information systems.

The Board of Professional Surveyors and Mappers (PSM) is another organization you can join. A licensed surveyor and mapper makes exact measurements and determines property boundaries. They provide data relevant to the shape, contour, gravitation, location, elevation or dimension of land or land features on or near the surface for engineering, mapmaking, mining, land elevation and construction.

Costs

Be prepared to make a significant investment if you are going to pursue this as a profession. A DJI S-800 multicopter with gimbal and camera will cost between $10,000 to $20,000 and you may need to add a second drone for reliability. A ground control station will allow you to see real time video and do some processing. If you want to fly and take pictures you will need GPS and a GPS hold function to fly to a position and hold so you can take photos. You will need $1000 for extra batteries, $1000 for a charger, spare propellers, etc. and $1000 for transportation cases. If you need flight training you will be spending $300 - $5000 per person per week. You will need a ground school and flight training and possibly a simulator. Software licenses are normally required for the autopilot. They have updates every few months and the cost is $0 to $4,000 per year. For Mosaic services you can pay $500 - $10,000 per year and professional association fees will cost you between $50 - $100 per association.

Labor

Labor will be your largest expense that you will want to take measures to minimize. You may hire someone to do marketing, do scheduling and take phone calls. Most jobs take two people but some require three. For example if you are in a boat, the boat driver, the drone pilot and the drone sensor operator. To save on costs, pay by the job vs. the hour to save on costs. You may also need a company to do post processing of the data like editing the film of a golf course. You will also need someone to do billing and bookkeeping.

Flight Consumables

Consumables fall into three major categories; gas, batteries and routine maintenance. You also need to consider travel costs. If you are local it won't be much. But if you have to travel, you need to consider drone transportation costs, hotels, rental cars and meals.

Figure 5 Drone Ground Control Station.
Courtesy of Troy Built Models

Communications

The FAA Nattional Policy Document N8900.227 says you will need an observer to look out for traffic while the sensor operator accomplishes the mission. The observer and pilot must have communications. Some options include cell phones, walkie talkies and a hand held nav/com receiver.

Maintenance Subcontracts

You can do preventative maintenance yourself or send back to the company. The time interval should be around 25 hours. This maintenance will take the drone out of service so that is why it's nice to have two. If you decide to start a drone maintenance company you will need to do a quick turnaround, be able to do gas and electric engine repairs, check for loose nuts and bolts, examine propellers, and inspect the electrical system.

Advertising

Some of the potential ways you can advertise your drone business is via a website, google adwords, magazines, networking and professional organizations, trade shows and referrals. You should build a website for your business application. Take a look at competitor websites. On your website you may want to show your least expensive packages and have customers call for quotes on expensive packages. Customers can call in to obtain pricing on the more expensive packages. Categorize your photos as industrial, mapping, agriculture, real estate, etc. Build a promotional video of yourself and your work so customers can gain confidence. Advertise in Google Ad Words. Magazines and Newspapers should not be used unless they are local and even then they are very expensive. Network through professional associations like LinkedIn or Craig's List. Join the Professional Photographers of America and any industry you have an interest in like boating, skiing or rowing. Attend photography or agriculture conventions. Go to a local golf course and offer your services for free. Then ask for referrals for other golf courses.You may want to consider LinkedIn (200 million members) and Craig's list

Insurance

You can expect to pay $2000 per year for your drone depending on size, weight and classification. You may want to consider the Unmanned Aircraft Professional Association (ua-pa.com) which provides legal, insurance and business counseling advice. You can select an application and ask for a business case analysis. You can select from a list of insurance services from the partner insurance firms that specialize in drone operations, services and manufacturing. You obtain legal services from the partner law firms that specialize in drones. When you take a job your customer may require up to $1 million of liability insurance. There are a handful of UAV insurance providers in the US. A list is shown in Appendix 13

Software Licenses and Services

A software license will be required if you desire to update the autopilot firmware. These updates range from free to around $4,000 per year. If you are doing a mapping application you can upload many photos to a mosaic company website and get one high resolution image back. Fees for a yearly subscription service range between $500 and $10,000 per year.

Schedule

Travel will be a major component of your schedule. Make sure you plan in delays. Don't forget that weather can be an issue and delay your work. You will need to organize your job around a processor company to determine when your results will be ready. Try to minimize your travel as much as possible. Have a plan for where to take the shots. Make sure you have spare batteries and a charger ready to go.

Data Processing

It's not the data, it's what you do with the data that will make your business successful. It's easy to take a photo or video but other sensors require sophisticated processing to obtain a profitable result. In order to process data from LIDAR, IR and hyperspectral images or even photographs for mapping or photogrammetry, you will need to make a choice. Purchase software and learn how to do your own processing or use an outsourced service. If you want to do it yourself, be prepared to make a significant investment of your time and money. If you want to do it yourself, one product you can buy is Pix4D Desktop 2.0 which will convert thousands of aerial mages into a 2D georeferenced mosaic or a 3D surface model with centimeter accuracy. The output file format can be read by ERSI, ArcGIS, Autodesk and AutoCAD. They offer their conversion service through the cloud. There is no price on the website so you will have to call for a quote.

If you want to outsource your data processing, one service provider for high resolution aerial images is Dronemapper.com. They make it easy for you to upload your pictures through a web based interface. Their website gives you instructions on how to take shots and plan your flight paths. Their pricing is determined by the number of square km of data you submit. I believe the processing segment will grow and more companies will be created to select services from. The key is to make sure the processor is compatible with your application.

Robata has a software product called Agisoft Photoscan Pro that can process thousands of images to produce photogrammic data. The final image is a high resolution, 5 cm accuracy detailed digital elevation model. A company called Altavan will use their aircraft to gather the data, organize and process it and deliver a final product to you.

Today the data processing is done off board at the ground control station or non-real time with recorded data. Low bandwidth availability causes problems displaying real-time full motion video. We will need high performance computers that have low power requirements and low weight. These new processors will need to be scalable with the sensor or suite of sensors. In the future, the operator will select the feature extraction needed, the processing will be done onboard the drone and the result will be streamed down for final decision making.

Menci provides photogrammertry software that uses small digital images acquired by drones to create a large map. This software runs on your PC and allows the user to create both 2D and 3D maps. You can import both .TIF and .JPG files. There are even three add on products. Stereo Tools will allow for 3D feature collection. Google Earth Tiler will allow you to share your image map through Google Earth. It builds KML files that overlay your image on Google Earth. Cloudview is a point cloud stream viewer. There is another software suite they are working on called KT-3D which does not require GPS or IMU data. You can visit the website at www.menci.com and download the software for a free trial.

FLIR offers a free App for Apple and Android that allows you to control your IR drone camera remotely through a WiFi connection. You can remotely adjust the temperature span and contrast levels, change color palettes, add temperature measurement tools, playback voice comments, focus manually or automatically. There is even support for multispectral images. You can take pictures or record live video on your mobile device.

FLIR also offers FLIR Tools for PC and Mac. This software will allow you to import, edit and analyze images and then turn them into professional inspection reports for your customer. You can directly import the images to your PC through a USB cable. Edit radiometric images to thermal tune level and span, change the palette, or adjust parameters such as emissivity, reflective temperature, and more. Add measurement tools like spots, area boxes, circles and lines. Create professional PDF image sheets and reports. Display stored compass and GPS information. Playback stored MP4 video imported from the camera. Switch between IR, DC, Thermal Fusion, PiP, and MSX modes.You can re-open saved PDFs for further editing and revisions.

Drone Internet Biz

Payload = Router

On September 2, 2013, a drone enthusiast decided to build a Wi-Fi hotspot drone that provides Wi-Fi over a local LTE network. It can even hook up to a VPN for more security. This could be useful after a natural disaster or just help people to get a good connection.

Satellite Internet

Congress charged NTIA with creating and maintaining the National Broadband Map to display internet availability. As you can see in Figure 6, a large portion of the USA is still without broadband internet services[5]. I was surprised to see California, which has a population north of 38 million is not very well connected. In 2011, the U.S. Census bureau reported that 71.7% of U.S. households have access to the internet[6]. Satellite internet is offered by several companies using radio frequency feeds where a signal is transmitted from an earth station to a satellite which acts as a relay to the subscriber. There are many severe problems with this method of delivery. Due to the satellite being in geosynchronous orbit at 22,236 miles, a time delay occurs which causes video and audio feeds to be choppy. Latency associated with satellite service has been an order of magnitude greater than wireline broadband technologies. A geosynchronous satellite orbiting the earth at a distance of 36,000 km has a round trip latency of about 500 milliseconds (ms). The necessary signaling between the set-top box and the satellite controller, to request assignment of a communication channel, can double this to over 1000 ms, which would preclude use of many latency-sensitive services such as voice or video. In contrast, the maximum average latency found in terrestrial technologies is less than 70 ms[7].

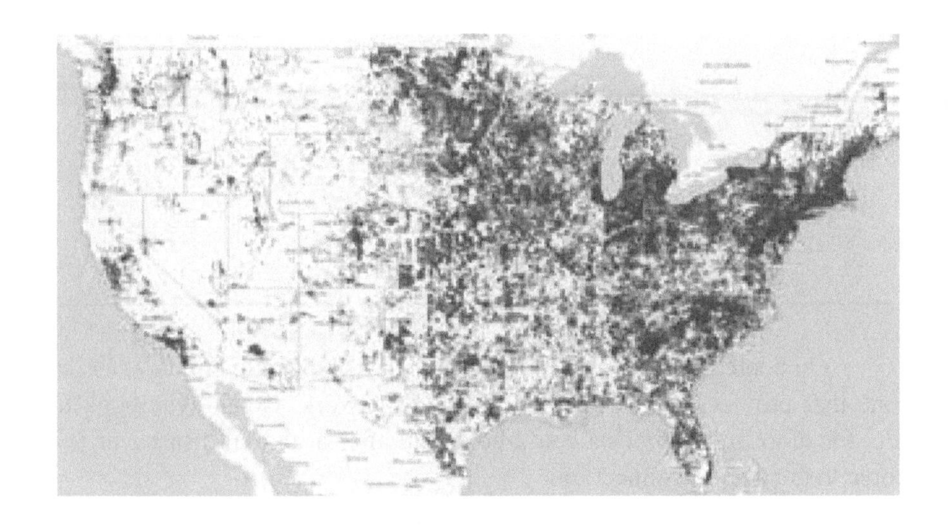

Figure 6 National Broadband Map with Speeds > 1.5 Mbps.
Courtesy of NTIA

Additionally, there is only a fixed amount of bandwidth available for all shared access subscribers. As a result, satellite internet providers have to place a low data transfer restriction on data over a one month period. If data downloaded exceeds some value, the subscriber is penalized with slower internet service so high usage subscribers will experience severe disruptions. In bad weather, reception will be attenuated and may completely shut down the link. Heavy rain and thunderstorms can completely block the signal. Snow can pile up on a parabolic dish antenna and disrupt the signal. Wind can bend a dish and disrupt the signal. And lastly, satellite internet is more costly than cable. Monthly cost depends on how much data is accessed and data transfer rates and ranges from $60 to $400. The average monthly service fees range from $86 (rural) to $163 (metro)[8]. In my research I have found that many are not happy with their satellite internet service or the outsourced technical support.

In order to provide drone internet service, it will be necessary to have long endurance. The U.S has a fleet of manned aircraft including the NOAA WP-3D and the USAF WC-130J that are used as hurricane hunters. An example our largest longest endurance drone is the NASA Global Hawk that is used for atmospheric research. Figure 7 shows a NASA Global Hawk above a hurricane ejecting a sensor package that drifts down into a hurricane via parachute. The package includes a GPS receiver and temperature, pressure and humidity sensors. Data is sent back to the Global Hawk and relayed to Dryden Flight Research Center. The Global Hawk can fly at 65,000 feet for 32 hours. It is in a class of drones called High Altitude, Long Endurance (HALE). Internal combustion engines (ICE) or heavy fuel engines are not suitable for drone internet service as they run out of fuel in a relatively short period of time. So we must use another method to design a drone propulsion system to extend the time aloft to a reasonable number for continuous internet service. There are many alternative power systems available that will keep a drone airborne for weeks, months and even years.

Figure 7 NASA Global Hawk above Hurricane with Embedded Sensor Package Ejection Photo. Photo Courtesy of NASA

Lighter Than Air (LTA)

Blimps and Zeppelins are in a drone class called Lighter than Air (LTA). This type of drone would be an easy, cheap solution to provide airborne internet service. A blimp is a non-rigid airship without a supporting framework. A zeppelin is an airship with a rigid metal skeleton. As a result, a zeppelin can be designed to be much larger than a Blimp and possibly carry a greater payload. Figure 8 shows a Navy airship called the MZ-3A. The Missile Defense agency awarded Lockheed Martin a $149.2 million dollar contract to build a high altitude airship. The concept is to deploy 10 airships to protect the coastlines of the U.S. LockMart is now working on a demonstrator that will have a one month endurance. The final production ready airship may be able to stay aloft for up to a year.

The weight of an alternative power energy storage system represents approximately one-third to one-half the total weight of the drone and payload. Possible energy storage methods include, chemical, electrochemical, electrical, mechanical and thermal. You will need to do a trade study on the optimal method for an airborne internet application to optimize the solution you are seeking.

Chemical Energy

One chemical energy method uses hydrogen. Hydrogen has about three times as much energy as gasoline. If a tank of gas in your car were to last 12 hours, an engine that runs on hydrogen would get 36 hours for the same weight. Boeing has successfully converted a truck Internal Combustion Engine (ICE) to run on hydrogen. They have integrated two of these engines into the Boeing Phantom Eye demonstrator which includes a large volume fuselage to carry hydrogen. It has a wingspan of 150 ft., flies at 65,000 ft., carries a payload of 450 pounds and has a maximum speed of 200 knots with an endurance of 4 days. A final production model may be able to carry a 2,000 pound payload and stay aloft for 10 days. Boeing had had several successful flights. They are currently marketing the Phantom Eye as a communications relay hub.

Figure 8 MZ-3A Airship. Photo Courtesy U.S. Navy

Solar Power

Another concept that has been successfully demonstrated is to replace the ICE with an electric engine. Demonstrated power sources for the electric engine are solar cells, batteries or fuel cells or some combination. The basic concept behind solar cells is to convert light to electrical energy. On a solar powered aircraft, the solar cells cover a given surface of the wing or other parts. During the day, power from sunlight supplies the current to operate the electric engine, electronics and charges the battery.

During night operations, the battery operates these elements. The available solar energy is a complex subject and depends on geographical area of operation, weather, energy collection and utilization, time of the year and time of the day, altitude and payload. A solar powered drone is composed of many subsystems which each exchange energy. Energy required is largely influenced by winds, aerodynamics and power use.[9] A solar powered aircraft design team must have experience in many fields including aerodynamics, actuators, sensors, electronics, energy storage, photovoltaics and their optimization.

While conducting research on solar powered aircraft I came across a very important paper from Andre Noth and Roland Siegwart called Design of Solar Powered Airplanes for Continuous Flight.[10] There is a section called power balance for level flight which describes an amazing concept. They start by using lift and drag equations and finally have a solution for power required for level flight. If this equation is integrated you can obtain energy required for level flight. The novel part is that electrical energy requirements were derived from aerodynamic equations.

Solar powered wing design is an area of research. Another amazing piece of work by H. Tennekes is called the Great Flight Diagram[11] and is shown in Figure 9. Tennekes did extensive research on birds and went on to develop a linear relationship between speed and weight and wing loading of anything that flies, from insects to 747's. Noth and Siegwart extended Tennekes research and developed empirical formulas from 415 sail planes and 62 solar planes that relate weight, wing surface area and aspect ratio. These formulas serve as a beginning for solar aircraft design.

Electrochemical Energy

Electrochemical energy storage methods include batteries and fuel cells. Current batteries use Nickel Cadmium (NiCad), Nickel Metal Hydride (NiMH) and Lithium-Ion. Li-Ion batteries are one of the most popular types of rechargeable batteries for portable electronics. A recent breakthrough occurred with Lithium-Sulfur (Li-S). Battery capacity can be obtained by multiplying watts x hours (WH) and specific energy is battery capacity/weight (WH/kg). Specifications for Li-Ion are 100-265 WH/kg and voltages of 3.7 or 4.2 volts. LiS batteries have demonstrated a specific energy of 350 WH/kg and are projected to achieve up to 600 WH/kg.

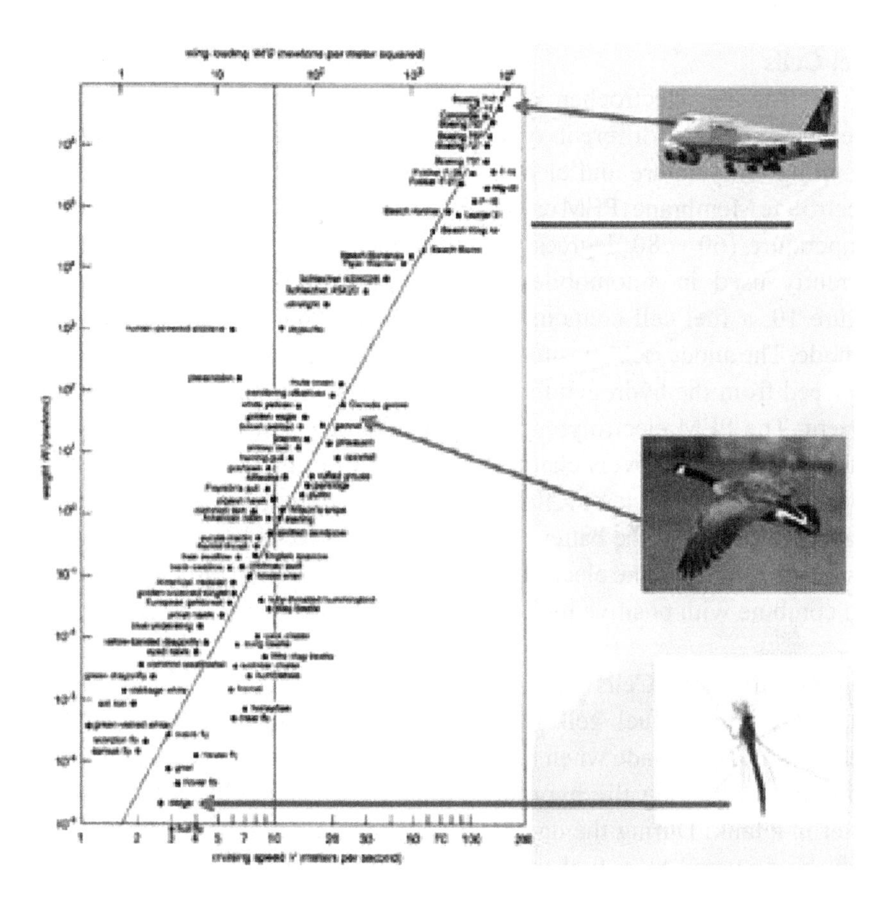

Figure 9 The Great Flight Diagram. Credit to H. Tennekes

The next generation of portable electronics is expected to run at 2 volts or less and Lithium Sulfer (LiS) cells have a voltage of 2.1 volts. Bottom line is that LiS could last 5-6 times longer than the current Li-Ion batteries and be compatible with the next generation of portable electronics.

Fuel Cells

Another electrochemical energy storage method is the use of fuel cells. There are several different types of fuel cells which are classified by their operating temperature and electrolyte. One of the most common is the Polymer Electrolyte Membrane (PEM or PEMFC). The PEMFC has a relatively low operating temperature (60 – 80 degrees C) and high power density. These fuel cells are currently used in automobiles that are powered by hydrogen. As shown in Figure 10, a fuel cell contains three basic elements[9]: the anode, electrolyte and cathode. The anode is the positive terminal of the battery and conducts electrons that are freed from the hydrogen molecules and flow in an external circuit generating current. The PEM electrolyte is like plastic wrapping paper that blocks electrons and only allows positively charged ions to flow through. The electrons are blocked by the PEM electrolyte and flow through an external circuit. The cathode is the negative terminal of the battery and has channels etched into it to allow oxygen to flow on the surface. The electrons that flow in the external circuit reach the cathode and combine with positive hydrogen ions to form water as a byproduct.

Regenerative Fuel Cells

If only a fuel cell is used, the time aloft will be limited. Another breakthrough was made when regenerative fuel cells were developed. The concept is instead of venting the byproduct water from fuel cell operation, capture the water in a tank. During the day an aircraft can be powered by solar power and at night be powered by a fuel cell. To increase endurance, the byproduct water is captured in a tank and at night, electrolysis is used to separate oxygen and hydrogen from water. These gasses are then pumped back into their respective holding tanks and the gasses can be re-used again and again. Regenerative fuel cells and solar power have the capability to keep drones aloft for years.

Recent Alternative Power Demonstrations

There are plenty of demonstrations that could be used to develop a business case. NASA has conducted an extensive multi-year research program on solar powered drones through the Pathfinder, Centurion and Helios programs. A picture of the Helios is shown in Figure 11. The Helios had a 246 ft. wingspan, 10 electric engines, lithium batteries and a fuel cell. The concept was to develop a high altitude aircraft to serve as a pseudo-satellite. Helios obtained a world record altitude of 96,863 feet.

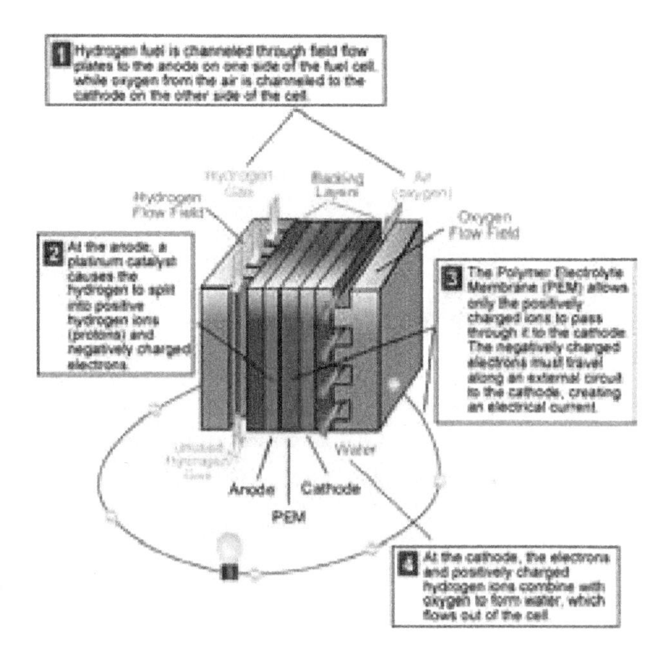

Figure 10 Fuel Cell Operation. Courtesy of the U.S. Department of Energy

On July 23, 2012, Quintiq used LiS batteries and solar power to keep the Zepher drone airborne for a world record 14 days. It used solar power during the day and batteries at night to keep electric engines and electrical systems running.

Silent Falcon Technologies has developed a small drone called Silent Falcon that is powered by solar power and batteries. This drone can stay aloft during the day (8 hours) + 6 hours at night on a Li-S battery for a total of 14 hours. It has a top speed of 70 mph.

On Oct 29, 2009, the Navy Research laboratory developed a fuel cell that could power a drone for 26 hours. The NRL Ion Tiger weighs approximately 37 pounds and carries a 4-5 pound payload. In Israel, the IAI small Birdeye drone flew with a fuel cell called the Horizon Aeropack H200 (Singapore). The normal battery powered drone had a 2 hour endurance. The fuel cell extended endurance to 6 hours. My research has shown that by substituting a fuel cell for a battery in a small drone, you can extend the endurance three to four times above a battery.

Figure 11 Helios Solar Powered Aircraft. Courtesy of NASA

Business Case

Your business case can follow two strategies. First you could focus on areas that lack coverage. Another strategy is to operate over a large city with millions of people and capture 10% of the existing market. Assume an average internet bill of $100 per month and your objective advantage as a startup is to lower internet costs. A single drone at 55,000 ft. has a coverage area of 314 km^2. Los Angeles city has a total population of 3.9 million and a land area of 1215 km^2. It will require 4 drones to obtain complete coverage. Annual revenues for internet sevice in LA are 3.9 million x $100 x 12 = $4.68 billion. If a drone internet startup captured 10% of the market, annual revenues would be $468 million. If you wanted to enhance the opportnity, you could cut internet bills in half. This would still leave you with $234 million in revenues. This is a huge opportunity that has significant potential for success.

Drone Crop Stress Biz

Payload = Optical & Infrared Sensors

The term precision agriculture has often been used in the drone industry, but I have found that few people really understand the meaning. I would offer the following as an explanation. Precision agriculture is the understanding of the complex interactions between crop growth and decision making. There are many chemicals used to protect and improve crop yields and these chemicals are generally applied over an entire farm. Using precision agriculture, you can identify problem areas and only apply enough chemicals to the affected areas. Instead of entire farm application, precision application will allow for significant savings.

In order to understand this business application for drone use, it is important to understand soil science, plant science and and sensor theory. Soil and Plant science is fairly complex and has many sub areas of study such as formation and classification and has physical, chemical, biological and fertility properties. For the purpose of this section we will focus on soil fertility.

Plants absorb more than 80% of incident sunlight in the 400 to 700 nm band[12] which means only 20% of sunlight light is reflected. In the near infrared band (NIR) of 700 nm to 950 nm, plants absorb a very low percentage which means NIR has a high reflectance. Figure 12 shows the spectrum for a singe plant types.[13]

The concept of evaluating the amount of incident light absorbed and reflected at different wavelengths has been utilized in the development of several ratios, collectively known as indices, which are sensitive to different environmental and physiological conditions. The Normalized Difference Vegetation Index (NDVI) is the most used and well known. NDVI only uses measurements from two sensors; optical and infrared. NDVI is the ratio of the difference between red and NIR bands divided by their sum. NDVI may be used to identify areas of poor soil fertility.[14]

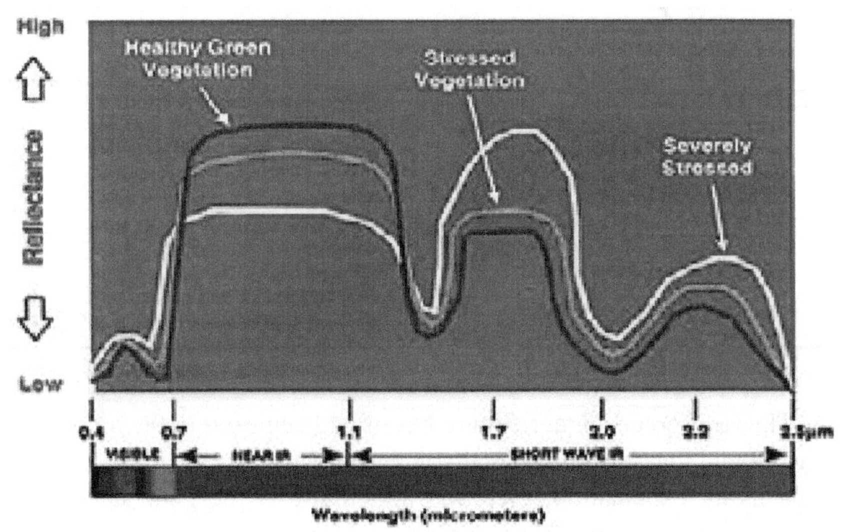

Figure 12: Plant Reflectance.
Courtesy of Federation of American Scientists

Currently, NDVI is calculated from satellite imagery. Farmers contact companies to produce NDVI images to make decisions on the health of their crops. Some advantages of using a drone equipped with optical and NIR sensors are that drones are significantly cheaper to use, there is relatively little time delay and the resolution is significantly better. With bad weather and non geosynchronous satellites, significant delays can occur in obtaining imagery. Significant delays can cause poor management decisions. Imagine making a decision on month old data from a satellite! There are actually off the shelf cameras (vegetation stress cameras) that can be mounted on a drone to take NDVI images Instead of making measurements and having to calculate NDVI, these cameras can take a picture and immediately display a NDVI image. One product is the XNite Cannon ELPH300ND, 12.1 megapixel vegetation stress camera. The satellite photo in Figure 13 shows an NDVI image for Phoenix.[15] Green is fertile and brown is where you will need to apply fertilizer.

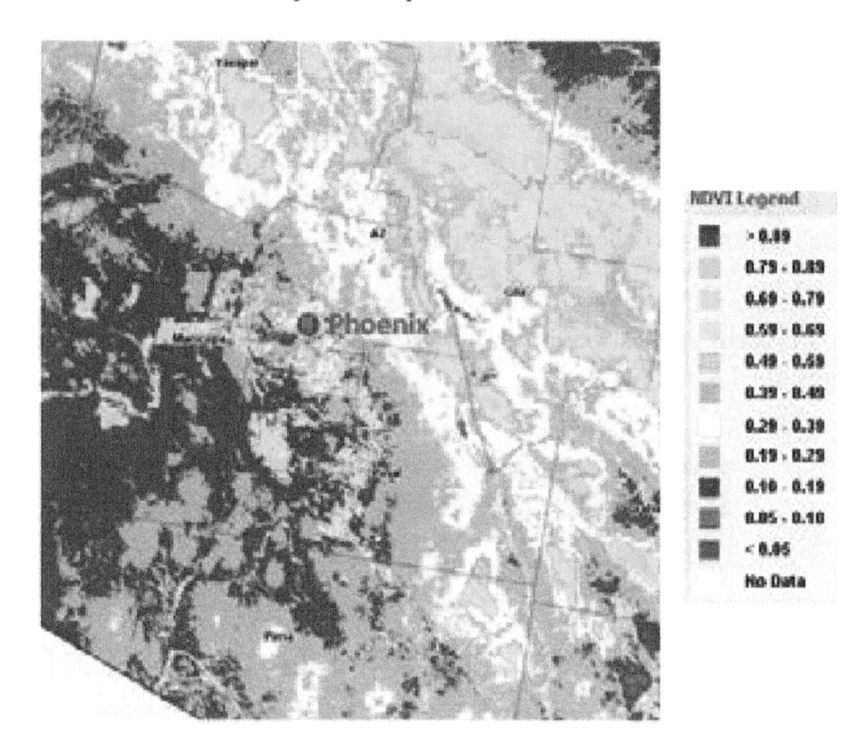

Figure 13 NDVI Image of Phoenix. Photo Courtesy of USDA

Business Case

This business has already been established. One North Dakota company, Field of View, works with farmers, agronomists and even a South American plantation manager. They believe there is a bright future in crop imagery using UAVs. Field of View has a product called GeoSnap which is an add on device for drone cameras that allows users to generate images with real world coordinates. Another company called Cropcam makes an easy to use drone to take GPS digital imagery. You simply upload the flight path, hand-throw the drone into the air and the vehicle will automatically skid land at the exact spot that it launched from. It is important to realize that in most commercial applications other sensors other than optical will be used and the rest of this book illustrates applications at other frequencies that are not visible to the human eye. How many customers are only going to pay for a picture? If you think you are going to mount a camera on a UAV and make a million dollares, you are wrong.

If we assume that a farmer uses fertilizer on his entire farm uniformly, our business case is to save on costs and quickly supply decision making data. The average farm in America is 480 acres.[16] If we assume that fertilizer costs $160 per acre, the cost to apply to the entire farm is $76,800. If we assume a modest 30% savings the total savings will be $23,040, a very significant number.

On a U.S wide basis, it can be seen from Figure 14 that a total of $26 billion is spent per year on fertilizer. A 30% savings in application would yield a savings of $7.8 billion.

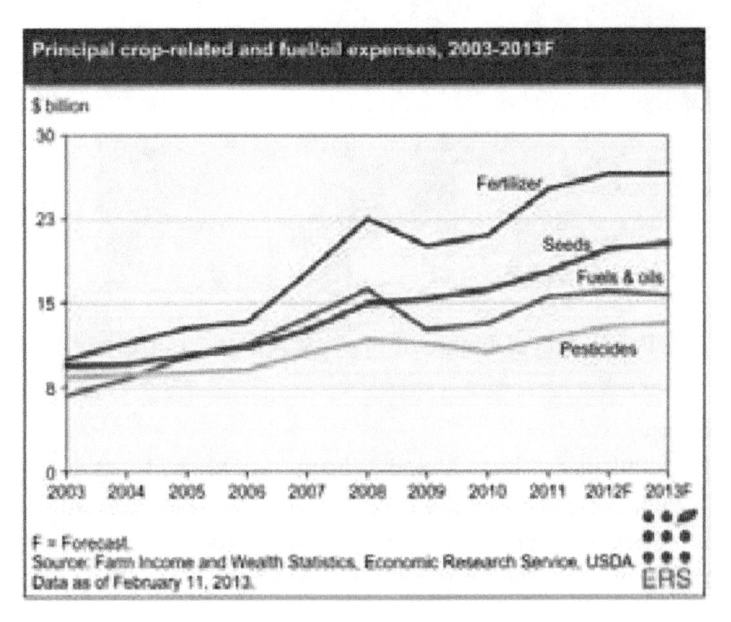

Figure 14 Cost of Fertilizer. Image Courtesy of the USDA

Drone Pathogen Biz

Payload = Hyperspectral Sensor

A pesticide is used to prevent, destroy, repel or reduce pests and the damage caused by pests[17]. Pests are living organisms that occur where they are not wanted and include insects, weeds and fungus.[18] Chemicals used to treat insects are called insecticides, for weeds, herbicides are used and for fungus, fungicides are used. Collectively these chemicals are called pesticides. Today, it is estimated that insects, diseases, weeds and pests eliminate half the food production in the world during the growing, transporting and storing of crops.[19]

Plant pathology is the science of plant disease caused by pathogens (infectious organisms). The categories for plant pathogens are viruses, bacteria, fungi, nematodes and parasitic higher plants. Nematodes are microscopic multi-cellular animals and parasitic higher plants are plants that cannot make their own food and become parasites to other plants. In this section we will focus on remote sensing for fungus detection.

Leaf rust is a fungus that affects wheat, barley, rye and grains. Once a fungus is inside of a plant, it absorbs nutrients and damages the plant, causing plant cells to rot. This plant disease has caused numerous epidemics and very high losses in the U.S. In the 1890's, most asparagus fields in the Atlantic States were entirely destroyed by stem rust.[20] In 1916, stem rust destroyed 38% of U.S. wheat production[21] Rust destroyed 25% of Texas spinach crop in 1937.[22] And in 1949-1950, mint losses in Oregon due to rust were 25-35%.[23]

Leaf rust is caused by a parasitic fungus called Puccinia recondita. An example of leaf rust infection in wheat is shown in Figure 15. Losses over large areas can range from 1 to 20%. Individual fields need to be destroyed when the disease is severe[24]

Figure 15 Leaf Rust. Courtesy of the USDA

Today there is a crisis in the coffee sector in Central America. Coffee leaf rust has reached epidemic proportions. As much as 70% of Central Americas coffee fields are affected. Farmers have lost hundreds of millions of dollars due to low production. Hundreds of thousands of jobs have been lost as there is less coffee to pick, process and export.[25]

Today, leaf rust inspection is done visually. Farmers observe the leaves and determine the amount of inspection by estimating attacked leaf area. Hyperspectral Imaging (HSI) offers high potential as a non-invasive diagnostic tool for plant disease detection. Depending on the interaction with the host tissue, pathogens cause disease-specific spectral signatures.[26] In a laboratory experiment, leaf rust infections were could be detected by a hyperspectral sensor data five days after inoculation. Early detection can prevent crop losss. A hyperspectral plot of sensor data is shown in Figure 16.[27]

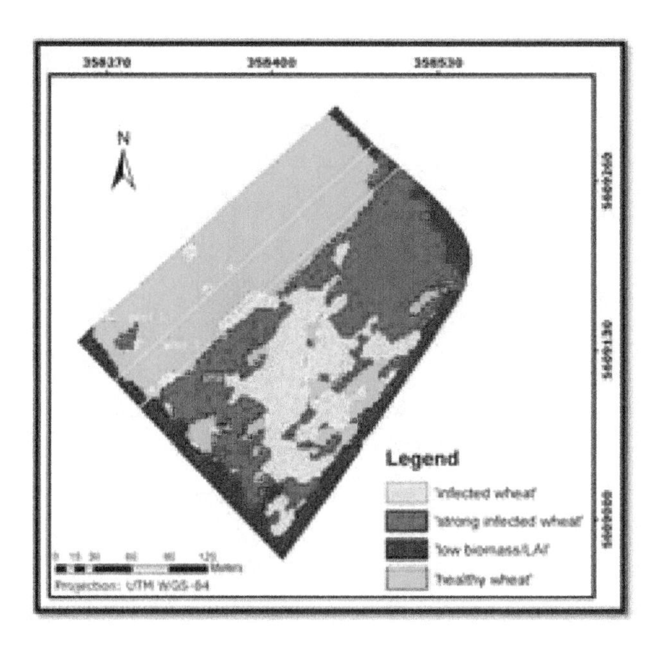

Figure 16 Hyperspectral Plot of Leaf Rust Infection.
Photo Courtesy of J. Franke

Once a fungus is detected it must be treated with fungicide. Fungicides are used to eliminate, reduce or remove the fungus. Fungicides actually limit the activity of the fungus and does not kill it. As a result, when the chemical wears off and the environment is favorable for fungus development, repeat applications will be required. This is good for a UAV business minded person as periodic inspection subscriptions could be sold for repeated business.

Citrus greening is one of the most serious plant diseases in the world. It is also known as Huanglongbing (HLB) or yellow dragon disease. This disease has devastated millions of acres of citrus crops in the US and around the world. The disease is spread by a disease infected insect, the Asian citrus psylid. Most infected trees die within a few years.

Figure 17 Citrus Greening. Courtesy of USDA

Symptoms commonly associated with the disease first appear as yellowing of the leaves. As the disease progresses, the trees suffer excessive leaf drop and foliage becomes sparse with fewer or smaller leaves. The disease also affects the fruit, causing it to ripen unevenly and become lopsided, visibly smaller and bitter-tasting. Once the host plant becomes infected, there is no cure for the disease. In areas of the world where the disease exists naturally, citrus trees decline and die within a few years and may never produce usable fruit. Citrus greening was first detected in the United States in Miami-Dade County, Fla., in 2005, and is only known to be present in the United States in the states of Florida and Georgia, the territories of Puerto Rico and the U.S. Virgin Islands, two parishes in Louisiana and two counties in South Carolina.

Both NIR and thermal IR can detect the presence of HLB.[28] Figure 18 shows a thermal imager response to HLB infection. Note that the left side is a normal image and the right side is a diseased image. These sensors can detect disease before the naked eye as is done now to hopefully intervene and cure with pesticide.

Business Case

Total U.S. expenditures for pesticides is shown is Figure 19.[29] If you add the totals for the three chemicals (insecticides, her-bicides and fungicides) you get approximately $13 billion annually in expenses for application of pesticides. If you calculate a conservative 30% savings by only applying pesticides where the drone remote sensor detects a problem, instead of the whole farm, annual savings would be $3.9 billion. I found one estimate for production costs for cabbage in Northern Florida. The estimate for fungicide expenses totaled $33 per acre, while herbicides cost $34.40 per acre, and insecticide costs were $199.33 per acre.[30] Using these costs, the average farmer with 480 acres would spend $266 per acre on pesticides which equals $127,680. If a 30% savings is realized the total saving is $38,304. A very significant number.

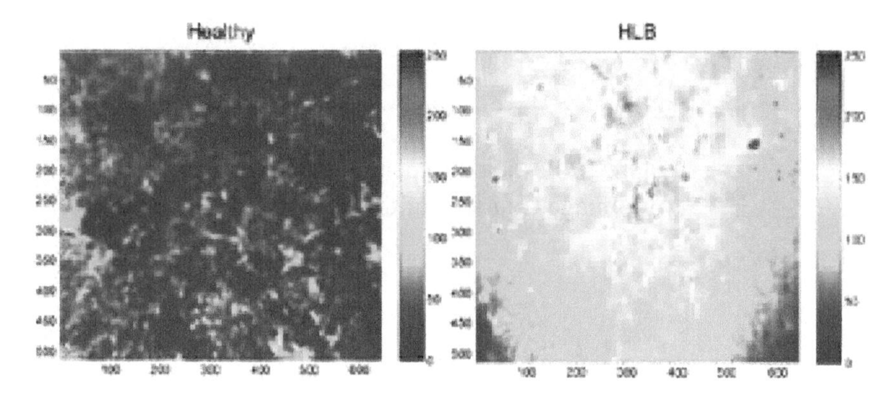

Figure 18 Thermal Images of HLB Infection.
Courtesy of the National Institutes of Health

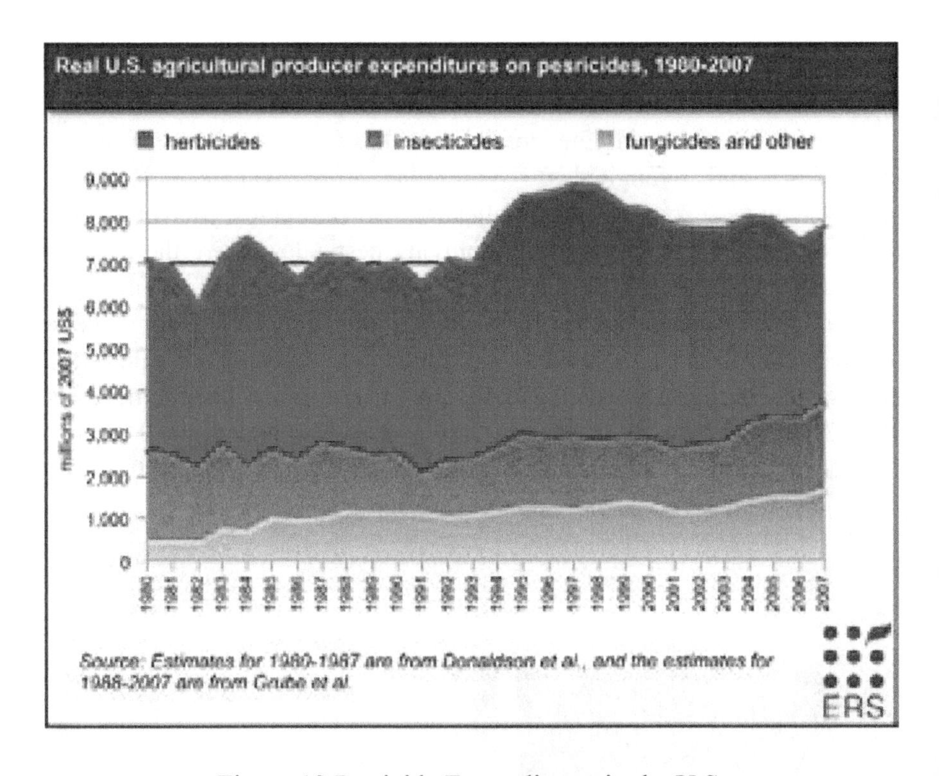

Figure 19 Pesticide Expenditures in the U.S.
Image Courtesy of the USDA

Drone Winery Biz

Payload = Thermal Sensor

How much irrigation water is required to grow quality wine grapes depends upon the site, the stage of vine growth, row spacing, size of the vine's canopy, and amount of rainfall occurring during the growing season.[31] Irrigation is critical to the harvesting of a successful crop and the major questions to be answered are when to start irrigation and how much water to apply. Measurement of moisture in the soil is the most common way to determine when to start irrigation. There are several methods used to measure soil moisture content and neutron meters (neutron probe) and dielectric sensors are preferred. A neutron meter uses tube inserted into the ground to a given depth with sensors inserted into the tube to make a measurement. Dielectric sensors measure the dielectric constant of the soil.

No matter which sensor you use, moisture sensors have issues. They have limited life expectancy, require frequent maintenance and are difficult to install. Because you are taking samples at discreet points, you are only getting an estimate of the entire vineyard.

One phenomenology associated with plants is that when there is a lack of water (water stress) the plant starts to heat up. It is possible to calculate a crop water stress index (CWSI) from the differences in temperature. A value of zero is no water stress and a value of 1 means maximum water stress. CWSI uses the difference between the plant temperature and air temperature and compares it to the dryness of the air (vapor pressure deficit (VPD). VPD can be calculated from the relative humidity. The end result is a plot that can be used to determine how water stressed a plant is. CWSI could be used to schedule irrigation. When the CWSI number gets to a certain level (say 0.6) it is time to irrigate the crops.

Figure 20 Thermal Image of Crops. Courtesy of USDA

You can start a business to measure the CWSI and use processing to formulate an image similar to Figure 20. Blue and green represent lower temperatures and yellow, orange and red represent higher temperatures.

Using a thermal sensor on a drone and a measurement of relative humidity will allow you to develop a product similar to Figure 20. Colors can be coordinated with CWSI values and when a threshold is reached, a recommendation for irrigation can be made to the farmer. This will allow for repeat business as your customers will want to know when to start watering. Additionally, the CWSI can reveal non-uniform watering patterns (where did you miss) and breakdowns in irrigation channels.

Drone Crop Dusting Biz

Payload = Fertilizer, Pesticle, Fungicide, Herbicide

There are 2500 unmanned helicopters being used in Japan for crop dusting. They have been in operation for over 20 years and now perform 90% of the crop dusting.[32] It is only recently that the US started to explore the usefulness of this application. Dr Ken Giles at UC-Davis is using the same helicopter, the Yamaha RMAX, to better understand how this vehicle can help farmers. A photo of the RMAX is shown in Figure 21.

Figure 21 Yamaha RMAX Helicopter. Photo Courtesy of Dr Ken Giles and UC-Davis

Some of the advantages I have discovered in my research are that a remotely controlled helicopter can be more precise in application. This allows for less over or under application increasing effectiveness. A helicopter can make tighter turns than a fixed wing aircraft and that will reduce spray time. For a normal fixed wing tractor aircraft, the chemicals drift down onto the top of the plant. The problem here is that the bugs and fungus normally live under plant leaves, not on top. The rotary motion of the helicopter blades allows for the chemicals to get under the plant leaves. Crop dusting is also done from ground based tractors and it can take up to an hour to apply to an acre of land. An unmanned helicopter can accomplish application in the same area in less than six minutes. After all, time is money.

Business Case

Because of the limited payload for the current generation of unmanned helicopters, I do not see these vehicles doing chemical application to an entire farm. However, if used to only apply limited amounts of chemicals to the areas that need it, there may be a significant savings realized. Certainly, labor cost will go down compared to ground based application. Crop yields could increase because neglected areas that are difficult to reach from the ground will receive application. And environmental impact will be reduced because exposure to chemicals will decrease. Spray drift, soil contamination, water pollution and health risks could decline. Rotary motion of the blades getting the chemicals where the bugs and fungus live could reduce the number of applications and as a result, realize significant savings. Crop dusting aircraft have been in many accidents and lives are lost due to running into power lines or poles.

The average cost to spray fungicide on one acre of farmland in Iowa is approximately $25 per acre.[33] Using the average farm acreage of 480, the total cost will be $12,000. I believe that in the future larger unmanned aircraft could do the same job as the fixed wing aircraft do today. Unmanning a crop duster will reduce costs and reduce risk to human life. There will be an upfront investment required but amortized over many years using depreciation will reduce annual cost.

Drone Weed Mapping Biz

Payload = Multi/Hyperspectral Sensors

A noxious weed is one that injures crops, habitats, livestock, ecosystems and/or humans through contact or ingestion. These plants grow aggressively and multiply quickly. There are non-native plants that were introduced into the country by ignorance, mismanagement or accident. Figure 22 is shows the extent of coverage of these weeds.

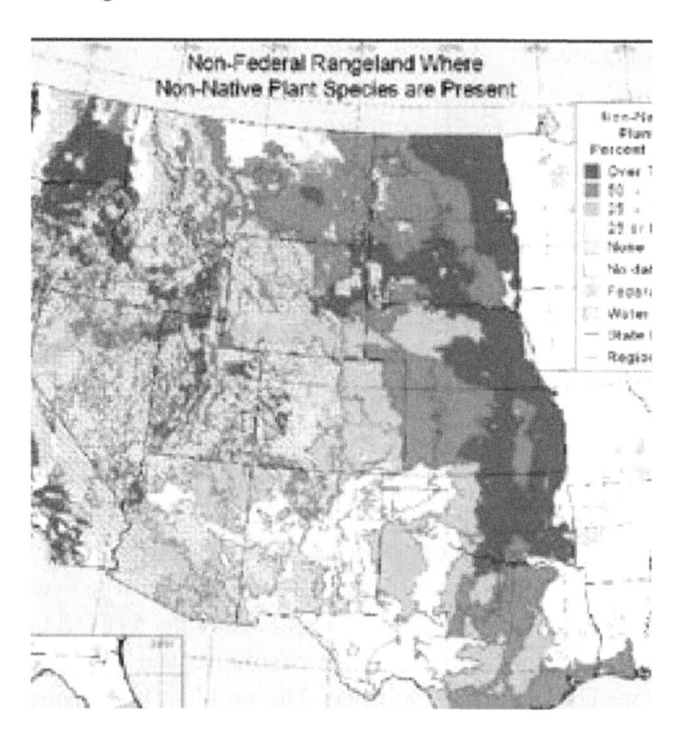

Figure 22 Non-Native Plant Species in Western USA
Courtesy of USDA

According to the USDA there are over 110 introduced, invasive and noxious plants in the USA. Non-native species are present in 50% or 1.7 million acres of the Nations non-Federal rangeland and cause losses estimated at $26 billion per year. Noxious weeds such as giant hogweed or poison hemlock can cause severe skin burns, sickness and even death. Several certain species have the ability to take over farmland and cause complete loss of crops. Livestock grazing is impacted, fire risk is increased, land is degraded and erosion occurs. According to the Bureau of Land Management, whorled milkweed was blamed as the cause of death of 19 horses in Colorado on Dec 15, 2012. If you are a farmer, it is important to develop a weed control program to prevent though early detection and eradication. The longer it takes to implement a control program, the more crop damage that occurs and the more expensive the control program will be. Aerial phonographs work best for weed detection when plants have unique growth patterns different from surrounding areas[34]. Drones can help with early weed detection

Each weed type has its own unique spectral signature and a drone with a sensor could make wide area assessments on farmland. Multispectral sensors can be used to detect and classify weeds by their unique spectral signature. A search of the literature revealed many projects that single out one type of weed using hyperspectral detection. In 1972, NASA launched the first mulitispectral imager called Landsat which provided optical and NIR sensors. They tried to accomplish weed detection but he resolution was too low.

In 1982, another satellite called the Landsat Thematic Mapper was launched. It contained an additional thermal sensor. The resolution was better but still in the tens of meters and only large dense patches of weeds were able to be detected. In 2001 NASA launched the Earth Observing-1 (EO-1) satelite which contained a hyperspectral sensor with 10 m resolution and detection of large area weed infestations could be accomplished. The problem with using satellites for remote sensing is of course the reduced resolution and the lack of availability of the satellites. Drones can operate very close to weeds, achieve centimeter resolution and can be deployed very quickly. Turnaround of the data can also be very fast.

Low Earth orbit (LEO) satellites rotate around the earth and periodically fly over the area of concern. If the weather is bad then "too bad, so sad", you will not obtain data and you will have to wait for the satellite on another day. Your imagery could be delayed for days, weeks or even months. Satellite imagery is also expensive for a small area and other jobs may take priority further delaying your request. Drones are available all the time, closer to the ground and may not be affected by weather.

A hyperspectral sensor that fits in your hand can now be carried on a small drone making it an affordable investment for someone that wants to start a business in weed mapping. This will be a business with annual subscriptions that will cover periodic inspections to prevent weed infestations. Imagery can be obtained immediately with high resolution centimeter accuracy. You can fly below the weather and there is a significant reduction in cost.

Drone sensors can obtain data from areas that are not accessible. The images could be fed into a geographic information database to monitor the spread over time. Once the affected area is treated, the effectiveness can also be monitored. You can contact the unmanned helicopter biz and ask them to discretely apply herbicide to the weed infested areas that you detect. The Rmax unmanned helicopter would be ideal for this. Figure 19 shows that the annual cost for application of herbicides on farms is $8 billion. If we can achieve a 30% savings using precise application by using drones for detection and application, the cost savings is $2.4 billion.

One problem we haven't considered is that herbicides used for many years eventually lose their effectiveness due to weed resistance that builds up over time. So you end up spending more for weed control. The typical numbers used to estimate costs are around $45 per acre. A recent USA report interviewed one farmer from Mississippi that said weed resistance increased his costs for weed control from $45 to $100 per acre. So if we choose a conservative $60 per acre the average farmer would spend $28,800 on weed control. A 30% savings in herbicide realized through precision weed control would equal $8,640

If we add up all of the savings from fertilizer ($23,040) plus pesticide ($38,304) plus herbicide ($8,640) that is a combined savings of $69,984. According to the USDA the average net cash farm income is forecast to be $89,000 for 2013. So a 30% savings using precision agriculture will almost double a farmers income. If the savings went to a more realistic 60%, a farmers income could triple. Figure 23 shows a summary of the savings by percentage. Maybe you start one or all five precision agriculture businesses but one thing is for sure, there will be an explosion in precision agriculture drone jobs.

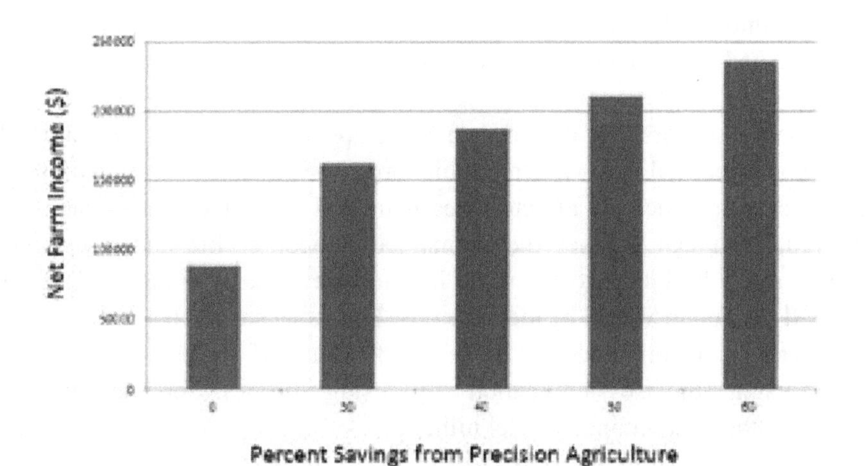

Figure 23 Net Farm Income Using Precision Agriculture

Drone Photography/Movie Biz

By TJ Diaz
Payload = Optical Sensors

This is a huge opportunity for anyone. It's fun and it's profitable. The average aerial photographer can make $300-$500 per hour. You can charge by the picture, charge by the flight or charge by the hour. Equipment can be purchased relatively cheaply. The best bet is to focus on mulicopters as they can take off vertically, hover and stare at something you want to photograph or make a movie of. You have less flexibility with a fixed wing drone as they require more airspace and cannot get into tight areas. The Parrot AR Drone 2.0 is only $299 and the DJI Phantom drone is $700. One recent success occurred when NBC hired the company Microdrones to make a film in Africa. The result was an incredible aerial film of African wildlife blended with some very entertaining music.

Photography vs. Cinematography

Your needs will vary greatly depending on whether you take photos or want to shoot high quality videos. Taking a photograph can be easily accomplished with a lightweight digital camera mounted on a fixed wing or multi-copter UAV. If you are working on high quality videos, a heavier video camera will be required. A large multi-copter is recommended. Cinematography is much less forgiving than photography. Your shots have to be smooth and your multicopter has to fly almost flawlessly to achieve desired results. The greater the number of engines, the greater camera weight you can carry. A custom designed octocopter is shown carrying a camera in Figure 24.

Figure 24 Custom Octocopter. Photo Courtesy of TJ Diaz

Stability and Control

Stability is an important factor in obtaining the clearest images. If your multi-copter isn't stable, you're photos will generally turn out poorly. The weight of your camera and lens will have a significant effect. Normally you choose the camera first and then find a UAV that can carry it. The next step is to determine the best way to stabilize your camera. You need to select a proper gimbal for your camera mount. There are a lot of gimbals on the market that you will need to select from.

The gimbal is usually the weakest link in the entire system. Gimbals typically offer pan tilt and roll stabilization. However, you must consider the form factor of your camera and then decide what size mount you will need for your application. In addition to the mechanical side of the gimbal, you will also need electronic gimbal stabilization; a system that keeps the horizon level. Gimbal stabilization is not cheap so budget will be a big factor in deciding on the system.

Here are some parameters to consider in the purchase of a small UAV:

Range
Endurance
Speed
FPV Telemetry & Waypoint Flying
Single Operator & Dual Operator Options

One advantage of unmanned aerial photography over manned aircraft is that a drone can get very low to take close-up shots. There are so many business opportunities including real estate, weddings, boat races, company building profiles, virtual tours, photogrammetry and many more. Even Hollywood is using drones to film the best selling films. Movies that have used drones to film include Star Trek: Into Darkness, Man of Steel and Iron Man 3. There is also a way to post your pictures inside of a social network called dronstagram. The website is still in beta phase but allows you to sign up and post your pictures in a number of categories.

Drone Search & Rescue Biz

Payload = Optical & Thermal Sensors

A fishing boat with 8 people on board 35 miles off Gloucester, MA starts taking in water. They are not exactly sure where they are but get a call off on the radio and a nearby ship intercepts. Within minutes, a swarm of drones from the coast guard is dispatched to the search area with optical and thermal sensors.

Figure 25 Bell Eagle Eye Drone. Courtesy of the US Coast Guard

The drones set up a low altitude gridded search area and initiate separate searches while communicating with each other. If one grid box is searched, then that box is communicated as covered and not searched again. The swarm searches the area in a quarter of the time it takes a single aircraft. When one of the drones comes in contact with the survivor, the position is communicated to a coast guard ship and airborne asset. Help is sent and the victim is rescued. This scenario is fictional but possible with the technology we have today.

Drone autonomy is a hot topic for research. The methodology is for drone sensors to be used as inputs to autonomous algorithms to enable drones to make decisions on their own. Drones have demonstrated they can fly in formation and communicate with each other and coordinate their actions. Realization of this scenario will occur in the not too distant future.

According to the US Coast Guard, there were 4515 boating accidents in 2012. There were 5900 vessels that caused 651 total deaths and 3000 injuries. Damages were assessed at over $38 million. Drones have advantages over manned aircraft during search and rescue operations including: rapid response, multiple sensors, long endurance and rapid distributed relay of victim's exact position with video.

In May 2013, the first rescue of a human from a UAV was accomplished by the Royal Canadian Mounted Police (RCMP) near Saskatoon. A young man was in a rollover car accident. Disoriented, he got out of the vehicle and wandered into the woods with very little clothing on a cold day. After a few hours he was very cold and called 911. The police were able to get a ping off the GPS in his phone but were not able to find him. So they called in a helicopter with night vision and a search light but were still unsuccessful. Then a RCMP officer trained to fly a drone called the Dragonflyer (see Figure 26) arrived.

He pulled the drone out of the car trunk, and launched it and within 5 minutes, The infrared sensor was able to find his heat signature as seen in Figure 27. When he was found, he was unconscious with hypothermia. The RCMP rushed him to the hospital and he made a full recovery. If the Dragonflyer had not found him, he would not have made it. This is a world first and as I stated in the Introduction, this is an example of the good things drones can do.

Figure 26 Dragonflyer Drone.
Courtesy of the Royal Mounted Canadian Police

In addition to this successful rescue, many other persons have been helped during hurricanes and earthquakes over the years. During Hurricane Katrina, the USAF made a request to the FAA to use drones to help but was denied. But during Hurricane Ike, the FAA approved use of the MQ-9 Reaper to help deploy rescuers to the most needed areas. Within one day after the Haiti Earthquake occurred, America sent its largest drone called the Global Hawk to find where the most damage occurred so rescue efforts could be concentrated.

In a number of cases aircraft have crashed in remote areas where they slice though trees and leave no indication of their presence. In the 1990s, NASA experimented with radar to use it as a search tool. In 1998, a test was performed in a swampy areas south and west of Virginia Beach. Aircraft parts were hidden in foliage. A P-3 aircraft equipped with a radar was able to detect the parts. A drone with a radar could accomplish the same.

In the UK, a drone called the AeroSee will transmit about 100 aerial images a minute and relay them to the Internet, where people who have downloaded special software can help search for a missing hiker. The images are streamed to computers and mobile devices. Users will be able to click on the image and create a tag if they think they may have discovered the missing person. The crowd sourced intelligence will be shared with rescuers who can assess the image and send the AeroSeek back to take a closer look.[35]

Business Case

There is an opportunity to start an on-call search and rescue business. You could charge an hourly rate for your services. Advertise your services with the police, fire department, forest service, etc. Obtain a contract.

Conduct free demonstration programs with a surrogate lost person, tape it and use for advertising. Attend local community events and develop a following as this is a local business opportunity and you will need community support. Create a contract for long term services and establish sub-contracts for on-call immediate services. One of the issues I have noticed about this business is that the rescue teams all try very hard to do a good job. But it is only after several days of failures that they call on the drone community. This is wrong as a drone should be a tool in the pocket of the search and rescue team that should be used immediately. Instead of using a drone as a last ditch effort, a drone should be used at the beginning of a search and rescue operation.

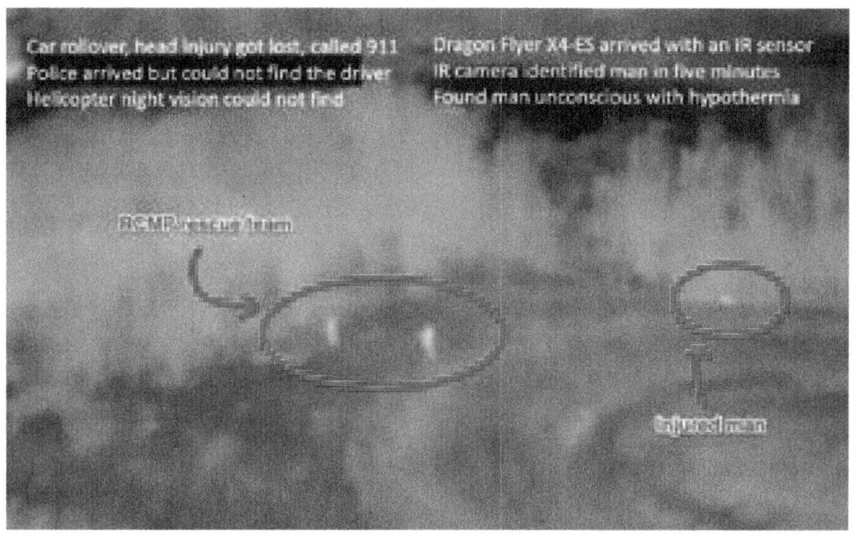

Figure 27 Drone Rescue with Infrared Sensor
Courtesy of the Royal Mounted Canadian Police

Drone Archaeology Biz

Payload = Thermal or Radar Sensors

"The only record of the history of our planet lies in the rocks beneath our feet: rocks and the landscape are the memory of the earth. [...] The record preserved in the rocks and landscape is unique, and much of it is surprisingly fragile. Today it is threatened more than ever. What is lost can never be recovered, and therefore there is an urgent need to understand and protect what remains of our common heritage."[36]

In October 2011, a small 35 ounce md4-200 drone from a company called Microdrones helped archeologists in Russia discover 200 burial mounds that were 2300 to 2800 years old. The data was used to create a digital elevation map and a 3-D model. Drones have been used in Peru, Australia, Spain and Turkey to allow archeologists see the 10,000 foot view. Findings have been published in the Journal of Archeological Science. Since 2009, Italy, France and the UK all have archeological research projects that use drones.

Remote sensing has been used in archaeology for years but it has been done from satellites. Moisture, color and texture variations of freshly plowed fields can reveal buried archaeology optically. Buried archaeological features affect how plants grow above them and can be revealed as crop/plant marks. Soil and shadow marks and snow, water and wind marks[37]. NIR sensors can be used to detect these features.

In the early 1980s the NASA Landsat was equipped with a thermal sensor and flew over Chaco Canyon, NM. Figure 28 shows a thermal IR image of over 200 miles of Prehistoric roads (900-1000 AD). The figure on the left shows the straight prehistoric roads. The figure on the right shows existing roads in yellow

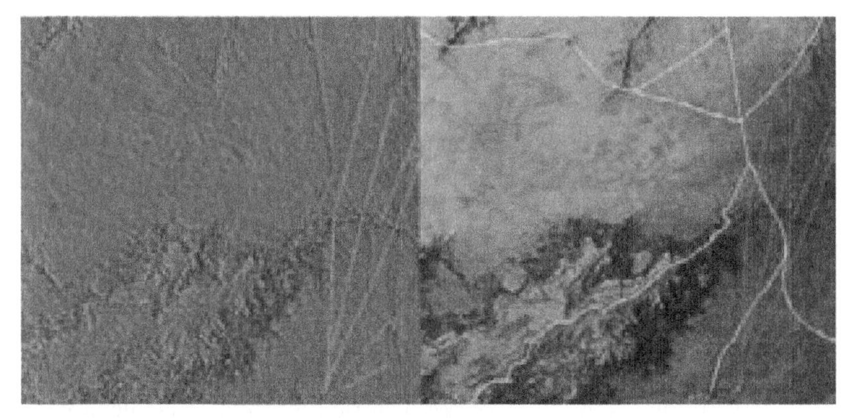

Figure 28 Landsat Photo of Prehistoric Roads Using a Thermal Sensor
Courtesy of NASA

overlaid on the prehistoric ones. AIRSAR is another NASA satellite equipped with radar. In 1998, evidence of a prehistoric civilization and remnants of 1000 ancient temples were found in Angkor Cambodia. This is considered the most extensive urban complex of the pre-industrial world. Angkor was built between the 8th and 13th centuries that is spread over 60 square miles of Northern Cambodia. The radar allowed detection of temples as shown in Figure 29.

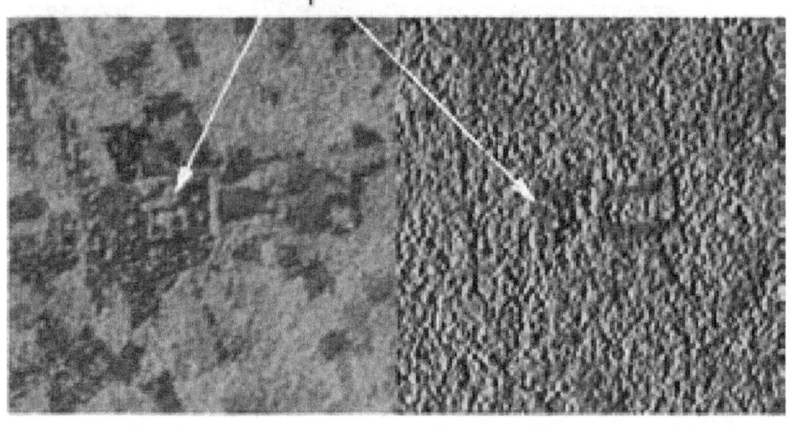

Figure 29 Temples Discovered in Angkor Cambodia Using Radar
Courtesy of NASA

Radars were normally too large to put on a small drone until just a few years ago when a company called IMSAR invented the world's smallest radar. The radar is the size of a pencil, is called the NanoSAR and can take real time radar images of the ground. The NanoSAR C can see at a distance of 1-3 km, weighs only 2.6 lb., measures 5.5" x 3.5" x 2", operates at X, Ku, UWB and UHF frequencies and operates in strip map, spotlight, circular and MTI modes. An advantage of radar is that is can see at night, in fog, rain, smoke, haze or snow.

Palentology is the study of pre-historic lifeforms. Palaeoichnology is a sub section that deals with fossil traces (invertebrates) and tracks (vertebrates). The main goal of palaeoichnolgy is the classification of the animal that made the track. Normally the study of traces and tracks are done with hand drawings, photographs or rubber casts. It takes a lot of time and the interpretation is very subjective. Figure 30 shows a 140 million year old dinosaur track taken in a quarry in Germany.

Figure 30 Dinosaur Tracks. Courtesy of USGS

USGS has planned a program to use a drone for high resolution photogrammetric mapping of dinosaur tracks in 2014.

Drone Food Delivery Biz

Payload = Food

In 2012 a website was created advertising Taco-copter a small business that lets you order by smartphone and have Mexican food delivered via a multicopter. The website caused quite a stir and many thought it was real but it turned out to be a concept and not a real service. Then came the burrito bomber. This food delivery machine turned out to be a real prototype. Using an app that can be downloaded, you order a burrito and have it delivered by a fixed wing drone. The drone has an integrated GPS and with some precision can pinpoint your location. Once it arrives, the drone releases the burrito that is contained in a paper tube and parachutes to the ground to the customer. Next came the OppiKoppi music festival in South Africa, an annual event featuring jazz and rock musicians where a multi-copter delivered beer from a drone via parachute.

As I read through all of these I really did not believe this would be a serious business application. But then I came across Yo Sushi, a UK based restaurant chain that is using a food tray (called the iTray) on a quadcopter to deliver sushi to its customers. The server controls the drone with an iPad and the cameras are fed back to the kitchen so the staff knows when to make more food. Service is expected to start in Fall 2013.

Finally, in the UK, where commercial drone operations are allowed, Domino's Pizza demonstrated pizza delivery with a multicopter they call the Domnicopter. A special container is attached below the copter and two pizzas can it inside. The operator launches the drone, it navigates to the customer and descends to a hand reachable level. The customer opens the container, reaches in and pulls out the pizzas, closes the container and the copter returns.

A Silicon Valley startup company called Matternet has a concept to create a network of drones to deliver food and medical supplies to remote areas with no roads in Africa. There are over one billion people in the world without access to all-season roads. Matternets' vision is to create a new paradigm (do you really need roads?) for transportation. Using small battery powered drones that have less than one hour of flight time, the concept is to have drones travel between relay stations where the batteries can be swapped out for fully charged ones. The company will deploy the initial hardware and maintenance services and set up a basic infrastructure network. In addition to a single drone with multiple stations, the company wants to create a network of multiple drones to expand the coverage area. Matternet has already partnered with an organization to help manage a drone fleet to deliver medical supplies in West Africa

In India they have had the worst monsoon flash flooding in 50 years in Uttarakhand which is sandwitched between Pakistan, Kashmir and Nepal. This disaster claimed over 1000 lives and 5100 are still missing. A total of 12,000 were rescued by the Army and Air Force. An India startup called Social Drones is using multicopters to survey the effects. Over ten other companies have started up drone companies to carry food and medical supplies to isolated victims.

Food delivery by drone is happening across the globe and it is easy for anyone to start a food delivery business. Delivery will occur faster than a car because there is no traffic. Less fuel will be used in a drone than the gas in a car. You don't have to pay someone to deliver the food item so you save on labor costs. And I believe someone will order food just to see the copter in action for the novelty of the experience as was validated by the email responses that Dominoes got after their demonstration. Currently they have over 1.4 million views on the YouTube video of their Domnicopter demonstration.

It wont be long until restaurants and fast food franchises start delivering food to your doorstep!

Drone Journalism Biz

Payload = Optical

In 2013 there were two schools teaching drone journalism; The University of Missouri School of Journalism and the University of Nebraska-Lincoln. The schools teach students to operate and experiment with drones to gather newsworthy data and conduct reporting. In addition, the students learn about FAA regulations and safety, how to take photos and how to integrate the information gathered into a story.

There is even a Society you can join: The Professional Society of Drone Journalists which offers updates on the development of all facets of drone journalism, teaches safety and regulations, teaches ethical standards and provides collaborative peer support. They advertise providing best practices for a variety of reporting including sports, local, disaster, weather and environmental reporting. A second society you can join is called the Professional Society of Drone Journalists (PSDJ). They claim to be the first international organization dedicated to the ethical, educational and technological aspects of journalism. Reporting categories are: investigative, disaster, weather, sports and environmental journalism.

Drone journalism is already very active in Australia as Asher Moses reports in the Sydney Morning Herald. Fox sports began using drones for coverage in 2012. The paparazzi have used drones since 2010 to cover Paris Hilton on the French Rivera. ABC journalist Mark Corcoran said "The day is fast approaching where a small personal drone will be an obligatory part of the tool box for journalists, photographers and bloggers" The 60 minutes program in Australia used a drone for investigative journalism after they were denied access to a detention center. They flew a drone overhead and captured imagery of a fire and overcrowding.

The laws for flying commercial drones differ by country. Transport Canada regulates Canadian airspace and requires that you obtain a special flight operations certificate if the drone is used for commercial operations. In Mexico, the government does not regulate and even encourages people to operate to make a profit. In the UK they have CAP 722 that regulates commercial drone operations. If the drone is less than 20 kg you only require a minor permit to fly. If the drone is heavier, it will require a permit to carry out aerial work. In Europe the European Aviation Safety Agency (EASA) agency is operating like the FAA only granting permits on a case by case basis. In Brazil there are no regulations for commercial drone use. Australia is the most supportive of all for commercial drone use. The Civil Aviation Safety Authority (CASA) only requires and easy to obtain permit. Finally, Thailand only requires permission of the landowner to fly over.

If you are a journalist, you can easily start a drone journalism business. If you are not a journalist, become a sub-contractor. Invest in a vehicle (most likely a multicopter), learn to fly it and start calling journalists and ask them if you can team with them in an on call basis. Write up a short contract with a daily rate and provide quotes for your service. Once you get your first customer, you are in business.

In the future, drones will be used on a routine basis to cover stories and sports events. Put it together and when the story is complete, land the drone, disassemble and move on to the next story. You can purchase a drone to get the right angle or shot to make the story more interesting and keep the viewers' attention. They will fly during the Olympics to collect the hottest videos of all events. If the Goodyear Blimp can do it, so can a drone.

Drone Mining Biz

Payload = Hypersectral

Mining companies around the world are trying to improve safety and efficiency. They are measuring conditions, inspecting pit walls, calculating quantities and making 3D maps. Hyperspectral sensors have been used on satellites for a long time for remote sensing. What is so special about hyperspectral sensors? They allow you to see a fourth dimension which reveals the chemical properties of materials. The imagery is formed by collecting data in different IR bands and then assigning a color to that data called a false color image. Figure 31 shows a hyperspectral image for Saline Valley, California.

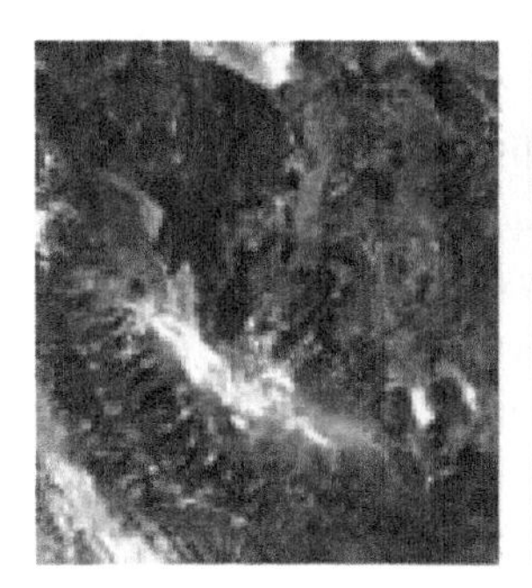 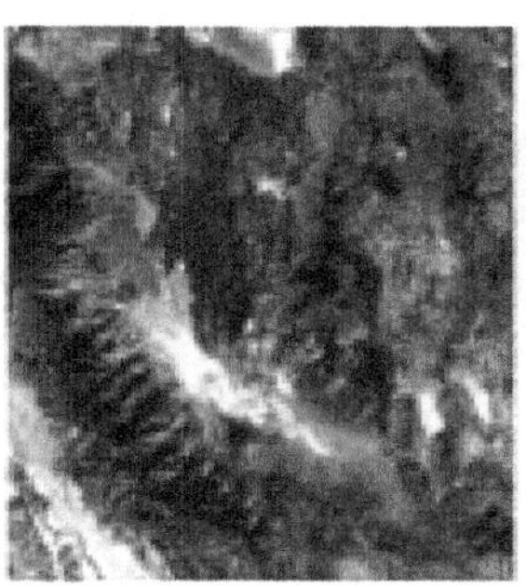 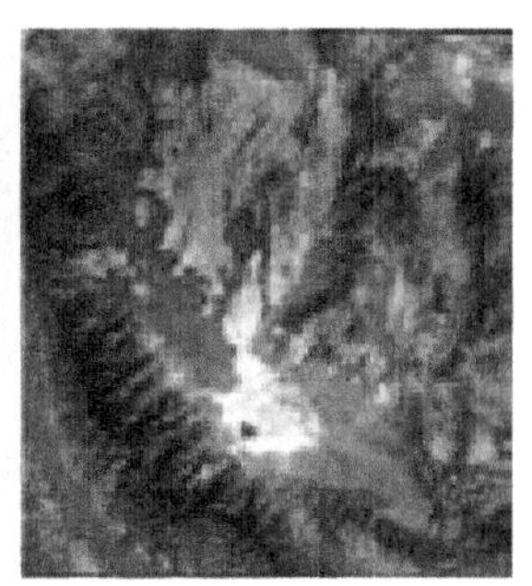

Figure 31 Hyperspectral Sensor Image for Mineral Exploration
Courtesy of the NASA/Japan Space Team

The left image displays visible and near infrared bands. Vegetation appears red, snow and dry lakes are white and exposed rocks are brown, gray, yellow and blue. Rock colors reflect the presence of iron minerals. The middle image displays short wavelength IR bands. Limestone's are yellow-green and purple areas are kaolinite rich. The right image displays thermal IR bands. Carbonite rocks are green and mafic volcanic rocks are purple.

Each mineral has a distinct spectrum. To automate measurements, if hyperspectral sensor spectra for each mineral were stored in a database, then measurements can be compared to the database and a match may be found. Figure 32 shows the distinct spectrums for a few minerals.

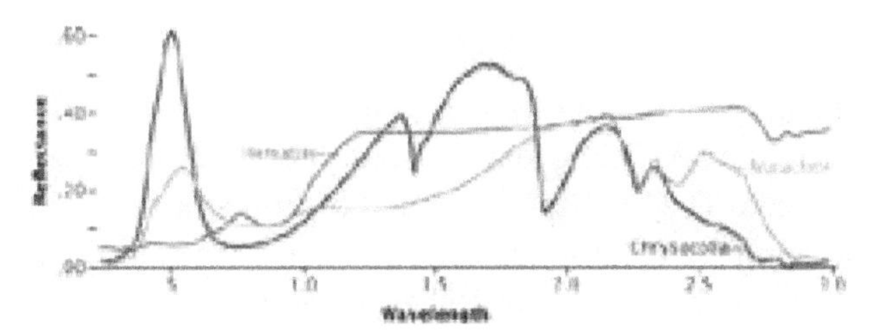

Figure 32 Hyperspectral Spectrum for Various Minerals.
Courtesy of NASA

You can start a mineral exploration business and offer your services to mineral exploration companies. You may want to join the National Mining Association (NMA) and network at tradeshows with samples of your work. Develop a portfolio you can take on the road. Draw up a contract and decide on hourly or daily rates to perform mineral exploitation services. Find a data processor that handles hyperspectral data and coordinate between your customer and the processor about the type of minerals required to be discovered. In the end, you want to provide your customer with a product that allows them to make a decision about conducting mining operations in the most suitable areas.

Or you could work for yourself with a goal of finding and purchasing or renting the land with mineral deposits you discover. Bring in some consultants, mine the ore and sell for a profit. Find some partners to put up the capitol after your first discovery. There is a gold prospector in the Yukon that is using drones. You too may strike it rich!

Drone Leasing Biz

Payload = Your Choice

On the medium to large scale drone side, Boeing says the leasing market may grow to $10 billion within 10 years. In the years 2007 and 2008 Boeing won leasing contracts worth $312 million to supply the Navy and Marine Corps. In 2009 Boeing also won a $250 million fee for service contract from the US Special Operations Command. Israel's Aeronautics Ltd and Israel Aerospace Industries Ltd are leasing drones to Dutch and Canadian militaries. The UK is leasing the Hermes UAV from a French company named Thales.

The price range for the medium to large drones can range from $1 million for a medium sized drone to $220 million for the Northrop Grumman Global Hawk that is used by NASA for atmospheric research missions. This does not include launchers, ground control stations, spares and multiple aircraft that are required for 24 X 7 missions. Boeing is leasing drones to government agencies and militaries. Their products include the Scan Eagle, A-160 Hummingbird and the Little Bird drones. The Navy uses them for pirate surveillance and Special Operations uses them too. On April 13, 2009, a leased Boeing Scan Eagle UAV was deployed off of the USS Bainbridge destroyer to help rescue Captain Phillips. He was captured by pirates that hijacked the Maersk Alabama cargo ship. The drones optical and infrared cameras were used to track the lifeboat holding Captain Phillips. He was rescued after Navy commandos shot and killed three pirates in the lifeboat.

On the small drone side, a company in the UK called service-drone.com rents drones by the camera size. For a LUMIX GH-3 you can expect to pay 2,000 euros per day. A Cannon EOS 5D MKIII will cost you 2,500 euros per day and the RED scarlet-X 4K will cost 3,000 euros. The pilot fee, insurance and technical equipment is included. An additional camera operator can be purchased for 500 euros per day.

There are several companies that already lease out their drones. A company in Germany called Rent-a-Drone will rent you a quadcopter for 190 euros per day. They use the Microdrones MD4-200 and MD4-1000 with goggles and tablet PCs for control. Another company in Japan called Secam will let you rent a quadcopter drone as part of a security solution. You pay a subscription fee of 5,000 yen per drone every month and when a break in at home or business occurs, the quadcopter activates and goes to the scene to get a view of what is happening. They will be launching the service in early 2014. If it is successful, they plan to launch in other countries.

Advanced Unmanned LLC has a custom designed unmanned jet powered aircraft that can fly at 45,000 ft. for 15 hours. Their pitch is that satellite imagery is out of date and they can provide fast, high quality images in real time anywhere in the country. They have EO/IR, hyperspectral, LIDAR, SAR and other sensors. They go on to say that the imagery will be delivered to you in real time. They also provide data processing services for large volumes of data.

With capital you could purchase a fleet of drones to lease to others. If customers want to fly on their own they will have to go through your training program Otherwise you will add on substantial fees for your pilots to fly the customers missions. Partner with data processing companies to be a one stop shop. Don't forget to get the proper insurance and do a background investigation of your customer. You don't want them to walk away with your drone so be careful and require a deposit up front. Make sure you have maintenance personnel or backup aircraft in case the drone has a failure.

On September 23, 2013, Boise State University revealed it is paying $1500 per hour to rent the RQ-16 T-Hawk made by Honeywell. The school is doing a joint research program with the Department of the Inerior on UAVs and firefighting

Analysts seem to believe the $10 billion number is conservative. They believe that the leasing market will grow exponentially. This market is called fee for service which can be a try before you buy option. Drones can be sold and company pilots can operate to accomplish the mission.

Drone Firefighting Biz

Payload = Near Infrared Sensor

The average number of fires in the USA on an annual basis is a shocking 60,000 - 80,000 which represents 3 - 10 million acres.[38] InciWeb.org will give you real time status information on every fire. Helicopters that need to be refueled every two hours are used to help fight fires. In 2007, NASA and the US Forrest Service conducted a series of experiments to demonstrate fire detection with the Autonomous Modular Scanner (AMS) flown on a modified Reaper drone called the Ikahana. The Ikahana is shown in Figure 33. The AMS contained a thermal infrared imaging sensor. The data was sent to the National Interagency Fire Center in Boise, Idaho and to firefighters in the field to help them direct their resources.

Figure 33 Ikahana Drone. Courtesy of NASA

Figure 34 shows the AMS sensor data plotted on a Google map for the Harris Fire in San Diego County on October 24, 2007. Active fire is shown in yellow and previously burned areas are shown in shades of dark red and purple. Unburned areas are shown in green hues.

Figure 34 AMS Sensor Data Plotted on Google
Courtesy of NASA and the US Forest Service

Forrest fires have become a major problem for certain areas of the planet creating serious problems for society, putting in serious danger and even worse, often destroying vegetation, wildlife and soil, causing significant ecological and economic losses not to mention the dire social consequences resulting from the destruction left in the wake.

On August 29, 2013, a Predator MQ-1 from the 163rd Wing of the California National Guard operating from the Victorville airport broadcasted information to firefighters in real time. They were fighting the 12 day old Rim Fire near Yosemite National Park. The Predator has a NIR sensor and was sending data directly to the fire commander. On that day, the Predator discovered a flare up that otherwise would have gone undetected. This data helps the incident commander make decisions about what resources to deploy and where. This was the longest sustained mission by a drone that broadcasted real time information to firefighters in real time. The Predator has helped out other times including 2009 when a NASA Predator helped the US Forest Service image a fire in the Angeles National Forest. In 2008 another drone helped out during a series of wildfires that took place from Lake Arrowhead in San Diego.

One example of the areas most affected by the fires are in Canada, where their mass extension of forests makes it a daunting task for fire crews, both land and air. In some parts of the United States it is exactly the same. California, for example, due to its warm dry climate, is one of the states hardest hit by wildfires. South America, Chile and Argentina are among the countries that suffer the most. Southern Europe, Spain, Italy and Greece have problems each summer. Bottom line is we need more data and the only way we are going to get it is through experimentation, data collection and analysis. The rest of the world is passing us by as Chile and China already have drones dedicated to fighting forest fires. Mexico is working with a company in Spain called Nitroflex to develop an unmanned plane that can deliver fire retardant.

Drone America offers the Aerial Scooper™ drone that weighs 1300 lb. with a wingspan of 28 ft. and an endurance of 5-15 hours using a gasoline powered internal combustion engine. They call it an unmanned fire suppression system (UFSS). It can deploy 2700 lb. of water and retardant to slow down and suppress and initial stage fire. They are planning to test in October 2013.

Drone Oil Rig Inspection Biz

Payload = Optical

The Deepwater Horizon oil spill was the largest accidental oil spill in history killing 11 people and discharging some 210 million gallons into the Gulf.[39] The cause was attributed to a Blowout Preventer (BOP).[40] A subsequent gas explosion ignited a fire and the oil rig sank to 5000 feet.

Figure 35 Deepwater Horizon. Courtesy of the US Coast Guard

Oil rigs are large structures with facilities to drill oil wells and extract oil and natural gas. The platform can be fixed to the ocean floor or may float. The US Minerals and Management Service reported 69 offshore deaths, 1349 injuries and 858 fires and explosions on offshore oil rigs in the Gulf of Mexico from 2001 to 2011

A gas flare stack serves to protect pipes from over pressurization. During oil extraction, the stacks burn off flammable gas released by pressure relief valves during an unplanned over pressurization. Flaring is the controlled burning of natural gas produced during oil and gas production. These stacks require detailed periodic inspections. Traditional inspections can require long and expensive shutdowns. In some cases flight by helicopters is restricted by narrow spaces or other obstacles. Flare stacks generate tremendous amounts of heat so fast flight control response is critical.

Figure 36 Oil Stack Flaring
Courtesy of the Environmental Protection Agency

Inspecting an oil rig can require long shutdowns for detailed inspections which cause risk to human life. Expensive helicopters have been used to perform the inspections in the past but they are not able to maneuver in confined spaces. If a done is used to perform the inspection several advantages can be realized. First a drone will be very cheap to operate and results can be obtained immediately though video and photographs. This media can be analyzed over time to do a detailed analysis. The inspection can be done while the asset is operating which saves significant production time costs.

A company in the UK called Cyberhawk performed the first offshore oil rig inspection in history with a drone. They have contracts with Shell, Exon, Mobil and ConocoPhillips and many other oil companies. A recent project with Shell realized a 4.7 million pound savings. The company primarily does aerial inspection surveys using remotely operated aerial vehicles (ROAV). Their clients are oil, gas, petrochemical and utility companies both onshore and offshore in the UK, Europe, Middle East and around the globe.

Cyberhawk is the perfect example of another trend that is starting. Examine your line of work and determine if you can save time or money. If so, then you can possibly transition your existing services to use a drone. If you join the Unmanned Aircraft Professional Association (UAPA) at ua-pa.com you can obtain a free business case analysis to help you start a successful business. Other benefits include access to insurance companies that insure drones and aviation law firms that specialize in drones. Choose an application from the 200 commercial applications for drones in Appendix 7. Then, after joining UAPA, ask for a business case analysis to determine who your competition is and your potential to profit at your location. More on UAPA is included at Appendix 8.

Drone Wind Turbine Biz

Payload = Videocamera, Thermal & Ultrasound

Wind turbines harness the power of the wind and use it to generate electricity. Simply stated, a wind turbine works the opposite of a fan. Instead of using electricity to make wind, like a fan, wind turbines use wind to make electricity. The energy in the wind turns two or three propeller-like blades around a rotor. The rotor is connected to the main shaft, which spins a generator to create electricity.[41]

Figure 37 Wind Turbine Farm.
Courtesy of US Department of Commerce.

This industry employs 670,000 to keep 225,000 wind turbines operating across the globe. In 2011, 23,640 new wind turbines were installed around the world. Statistics show continued expansion with annual market growth of almost 10%, and cumulative capacity growth of about 19%. The capacity around the world at the end of 2012 was 282,587 MW and is forecast to be at 460,000 MW by 2015. The USA has 29% of the global market followed by China (28%) and Germany (5%), India (5%) and the UK (4%). Spain has 60% of total power supplied by wind turbines.[42]

The weight of a single wind turbine exceeds 150 tons and costs between \$2 and \$3 million. The GE 1.5S generates 1.5 MW of electricity and stands 100 ft. tall. The blades are 100 ft. long, sweep an area of 42,000 ft^2 and have a maximum blade tip speed of 183 mph. Gearbox failure is one of the most serious causes of failure because the cost comprises 10% of the overall cost of a single wind turbine. Gearbox failure causes two to three times more downtime than any other component failure and takes a week to replace. Over the 20 year life cycle one can expect to replace the gearbox two to three times. Causes are failure of the bearings, abrasion and adhesion and failure of the gears caused by corrosion and bending fatigue.

Figure 38 Wind Turbine Gearbox Failure.
Courtesy of Chief Gary Bowker

It has been observed that before failure the gearbox will heat up and over time that heat will not be able to be dissipated any longer. When this occurs, the oil catches on fire and structural failure occurs as shown in Figure 38.

So how can a drone contribute to preventative maintenance so this failure mode does not occur? A thermal infrared sensor like the FLIR Tau detects infrared thermal energy and converts it to an electronic signal which is processed to produce a false color thermal image with a range of temperatures assigned to each color. For preventative maintenance, a drone could climb up to the gearbox with a thermal IR sensor and measure the heat at the gearbox. There will actually be a temperature distribution because the gearbox is so big. The GE 1.5S gearbox diameter is 231 feet! After a number of inspections are performed you will be able to gauge what normal operational temperatures are so that you can detect an overheat condition and recommend replacement. You may even be able to prepare charts with different temperatures for different models. Eventually you could publish a book on all your measurements and observations. Something that has never been done before. Figure 39 shows the heat distribution for a gearbox.

Figure 39 Heat Distribution for a Wind Turbine Gearbox
Courtesy of FLIR

Statistics indicate that as more wind turbines are built, more accidents occur. From 2008 through 2013, there were 1500 wind turbine accidents in the UK alone. Since 2000 there have been 112 fatalities including the worst one in 2012 in Brazil, where a falling blade killed 17 bus passengers. By far the biggest number of accidents found were due to blade failure. Blade failure can arise from a number of possible sources, and results in either whole blades or pieces of blade being thrown from the turbine. Pieces of blades are documented as travelling up to one mile. In Germany, blade pieces have gone through roofs and walls of nearby buildings.[43]

To prevent accidents caused by blade failure, periodic visual inspections must be performed. Inspections are done visually from the ground using a telescope or climbing the tower to discover cracks or delaminations. These defects can occur due to strong winds, bird hits or even normal operational wear and tear. Climbing one of these gigantic structures is very dangerous. I have examined the accidents and found numerous fatalities caused by falling, electrocution by nearby power lines and one incident where ice fell from the tower and cut a person in half. Drones will reduce risk to human lives.

A drone could easily accomplish the optical inspection with a camera in a very short time with reduced risk to human life. But what if we could go one step farther and detect failures before they become visual. Wind turbine blades are constructed using fiber composites. These composites are much stronger than steel (up to 3X) and make up 50% of the 787 and A350 airliners. A composite consists of two or more elements which consist of fiber and a binder. Advanced composites are based on carbon fibers, carbon nanotubes and graphene. Wind energy is growing so fast in the world that by 2020 it will occupy nearly 60% of the composites market.[44] Delamination is one of the most common defects in carbon fiber reinforced plastic (CFRP) components, such as those used in aircraft and wind turbine blades. Delamination occurs when the layers of composite material begin to separate and lose mechanical strength. To detect delaminations, different nondestructive testing (NDT) methods such as ultrasonic (UT), eddy current (EC) scanning, flash thermography, and recently developed pulsed-eddy-current (PEC) simulated thermography.[45]

Use of these sensors would require a custom mount, data link and signal processing. An ultrasonic sensor on a drone for NDT has already been demonstrated by a Masters student in the Netherlands at the University of Twente. The student thesis shows actual design dimensions. The student performed a successful simulation, design and demonstration. The propulsion system used a lift fan on a quadcopter with a robotic arm and electromechanical manipulator. A successfull demonstration was performed with an ultrasonic sensor on a drone making contact with a wall. To balance the contact a counterweight was used on the side opposite the sensor.[46]

Now you have three services you can offer; gearbox failure prevention, visual blade inspection and blade NDT for failure prevention. Put a design together and get permission for a wind farm to test it or go to a wind turbine manufacturer. With the growth in this industry you will have a job for the rest of your life!

Drone Transmission Line Biz

Payload = Optical, Infrared and Ultraviolet

On November 15, 2012 a helicopter inspecting power lines crashed in Corning, NY. The copter was a MD 369D and belonged to Haver ield Aviation, a leading provider of power line inspection and maintenance services. The cause was clipping a power line. On August 5, 2013 two men from the same company died when a cable carrying them came in contact with a power line and snapped. They were inspecting a new power line.

Figure 40 Transmission Line. Courtesy of the Department of Energy

Lots of the US grid infrastructure was built in the 1960's and is failing. The major problems are conductors, insulators, structures, metal fatigue and cracking, metal corrosion, deterioration of line splices and right of ways (strips of land purchased by the power company). Unless these problems are fixed, large power outages will occur. Humans can visually detect defective components, encroachments, vines, and dangerous trees, Problems are usually documented with a camera or video.

The objective is to detect potential failures before they occur. Thermal infrared images can reveal problems. A bad or corroded connection will cause the resistance to increase which causes the temperature to increase. Figure 41 shows an example of this problem. If the connector is not fixed, it could break and result in a loss of power. A small UAV could easily be equipped with an EO/IR sensor to do optical and thermal IR inspections. There were 372,340 miles of transmission lines in the US in 2009 and the number is expected to grow to 406,730 by 2019. Enough for small business to survive on forever.

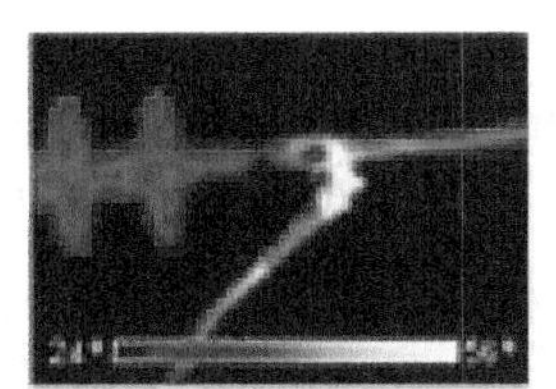
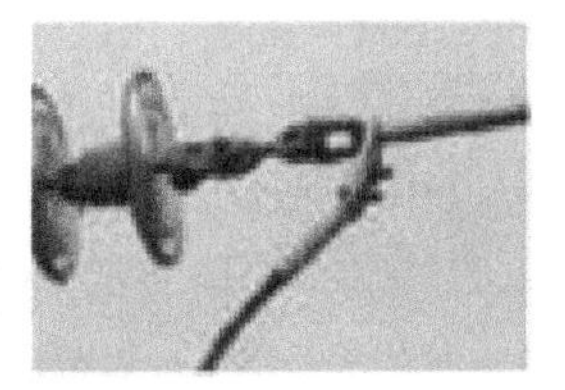

Figure 41 Thermal Image of a Defective Transmission Line Connector
Courtesy of FLIR

One of the issues you will have to address is proper calibration and shielding. There is electromagnetic energy surrounding power lines so you will need to calibrate and shield your drone to get accurate measurements. One major advantage is that you will be able to get much closer to what you are examining than a helicopter can. Helicopters normally fly 20 -80 feet above the lines. A drone will be quicker, safer and more cost effective than a helicopter. You will also have access in highly forested areas where helicopters cannot fly.

Since there are a large number of helicopters already in this market, you will need to steal a small share away for yourself. Maybe you can team with a company that just does power line repairs and add your subcontracted aerial inspection business to their services. Helicopters are expensive to maintain and operate. First you have to purchase one or rent, then you have to pay a pilot, observer and maintainer.

Normally you will be inspecting a given distance for the power line with a fixed number of towers. I recommend you do an analysis for what it costs for people to climb the poles and inspect and then do a second analysis for what it costs to perform the same inspection with a helicopter. Then you will have three columns to compare costs and the drone will win out for sure. Helicopter operating costs are $450 to $750 per hour. You could develop a pricing model by hour, by distance or by the pole and add an assortment of services like optical and/or infrared inspection, recording, documentation and reporting.

Power lines have insulators that connect them to utility poles and transmission towers to prevent short circuiting to ground or creating a shock hazard. When these insulators break down due to the corona effect, an ultraviolet sensor can see this discharge. Coronal discharge is undesirable in transmissions lines because it causes power loss, audible noise, electromagnetic interference, ozone production and insulator damage.

Figure 42 Ultaviolet Image of Coronal Discharge on Power Line Insulator. Courtesy of Nitromethane

Drone Bridge Inspection

Payload = Optical and Thermal

The transportation infrastructure in America is a ticking time bomb. The average bridge is 42 years old and over 170,000 bridges are risky to travel on. The National Bridge Inventory (NBI) is a database compiled by the Federal Highway Administration (FHA). It uses a rating scale that evaluates the condition of bridges:

Figure 43 Golden Gate Bridge. Courtesy of NASA

- 9 Superior to present desirable criteria
- 8 Equal to present desirable criteria
- 7 Better than present minimum criteria
- 6 Equal to present minimum criteria
- 5 Somewhat better than minimum adequacy, tolerate being left in place as is
- 4 Meets minimum tolerable limits to be left in place as is
- 3 Basically intolerable requiring high priority of corrective action
- 2 Basically intolerable requiring high priority of replacement
- 1 This value of rating code not used
- 0 Bridge closed

Additional classifications include structurally deficient, functionally obsolete and fracture critical. Structurally deficient means the bridge has a significant defect which causes speed or weight limits. A structure evaluation of 4 or lower qualifies a bridge as structurally deficient. Functionally obsolete means the design is not suitable for use. Fracture critical bridges, lack redundancy, which means that in the event of a steel member's failure there is no path for the transfer of the weight being supported by that member to hold up the bridge. These bridges are vulnerable to collisions with ships, large trucks or earthquakes.

According to the FHA, there are 607,380 bridges in the US and 67,000 or 11% (1 in 10) are classified as structurally deficient, 87,748 classified as functionally obsolete and 18,000 classified as fracture critical. In May 2013, the Skagit River Bridge in Mount Vernon, Washington collapsed when it was struck by an overweight truck. It was on both the functionally obsolete and fracture critical lists.

Figure 44 I-35 Bridge Collapse in Minneapolis.
Courtesy of the University of Minnesota

In 2007, the I-35 Mississippi River Bridge in Minneapolis collapsed killing 13 people and injuring 145 (Figure 44). This bridge was on the fracture critical list. Some of our most famous bridges on the fracture critical list include: Brooklyn Bridge (NY), Coronado Bridge (San Diego), Fremont Bridge (Portland) and the Lafayette Bridge in (St Paul).

Traditional bridge inspection methods are over 100 years old and include visual and audible inspections like hammer sounding, chain dragging and listening under load. Hammer sounding is one of the oldest and most commonly used methods for inspection. A dull sound or loud pop indicates a spalls or delaminations. A delamination is a separation of the concrete above and below the reinforcing bars (rebar) caused by corrosion. Delaminations cause a rapid deterioration and if they go undiscovered will eventually cause he entire deck to be replaced. Hammer sounding can also expose loose fasteners. For chain dragging, the inspector drags a heavy chain over the bridge surface. The chain contact produces an audible indication of delaminated areas. The areas are marked for further evaluation.[47] All of this takes time and considerable effort and is subjective to interpretation. Reliance on a single individual is subject to errors.

A variety of sensors have already been proven to be useful for bridge inspection. Ground penetrating radar (GPR) can find delaminations, see if the rebar is deteriorating and determine deck thickness. A GPR can be purchased as an option for the Schiebel S-100 VTOL drone. Thermal IR sensors can easily detect delaminations. LIDAR has been demonstrated with high resolution imagery

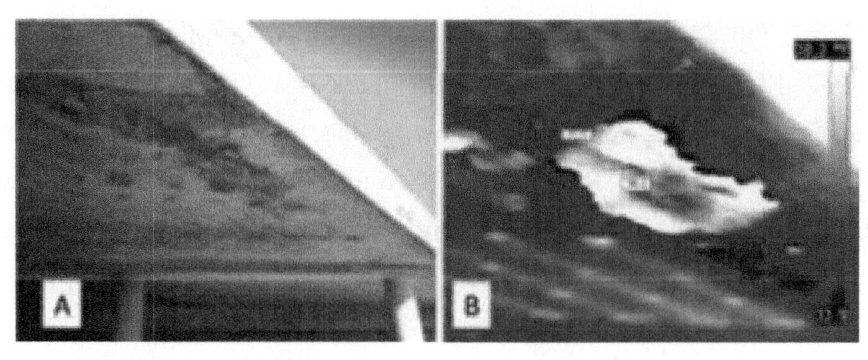

Figure 45 Thermal Sensor Detection of Bridge Delamination

to detect deck joint deterioration.

Sensors can be installed on a drone to aid in bridge inspections and more reliably detect faults and failures. The business case you need to make could be focused on a comparison of labor costs. Do an analysis to determine how much the city pays for bridge inspection. Then compare this to using a drone to perform the same inspection.

Figure 46 Hammer Pounding vs, Thermal Sensor
Courtesy of the Federal Highway Administration

Drone Roof Inspection Biz

Payload = Optical & Thermal

The average lifespan for a roof depends on the materials. A dark roof absorbs more heat and shortens its lifespan, The higher the pitch of the roof, the longer it lasts. A poorly ventilated roof shortens its lifespan. Tree branches rubbing and acidity from leaves will shorten life. Severe weather and with large temperature swings shorten lifespan due to expansion and contraction.

During the day the sun heats a roof. At night that retained heat is radiated back. Wet areas on a roof will retain the heat longer and dry areas will cool more rapidly. A thermal IR sensor can detect the uneven heat dissipation and identify areas of concern. Thermography has been used for roof maintenance using hand held thermal sensors. This sensor can be used to discover moisture or water damage, roof leaks, window leaks and heat loss. Using this sensor on a drone will allow you to quickly identify problem areas, reduce energy bills (deficient insulation) and reduce unnecessary and costly repairs.

Concentrate on the areas of the country that receive a lot of rain for roof leaks and focus on the cold northeast for thermal problems from lack of insulation. Thermal sensors are currently available for drones. So this is an easy one for you to make a business out of. Teaming with a home inspector could be a great way to get started. The home inspector can add on your costs and offer an additional service. With expensive multi-million dollar homes, this should be a shoe in. One small drone business in Florida is saturated with roof inspection requests from insurance companies.

Figue 47 Thermal Roof Inspection
Defective Leak Areas Shon in Yellow

Drone Anti-Poaching Biz

Payload = Optical & Thermal

Many countries are experiencing massive animal losses due to poaching. In Mozambique, every single rhino has been killed. In South Africa the numbers are 334 killed in 2011, 668 killed in 2012 and 800 rhinos killed in 2013. It is estimated that there are only 20,000 rhinos left. The same animal is being poached in India. Over the past year, 20,000 elephants have been killed in Africa as well. South Africa has let a request for proposal to have drones catch the poachers in the act. If we dont stop poaching, rhinos and elephants could become extinct in our lifetime. India has already started a similar program.

Figure 48 Rhinosaurous Tusk
Courtesy of San Diego Zoo

Rhinosaurous horns sell for $30,000 a pound on the black market making them more valuable than gold. In contrast, ivory from elephant tusks sell for $1,000 a pound. The illegal trade in animal parts is up to $7-10 billion a year.

In May, 2013, Russia started using drones to prevent poaching of brown bears, reindeer and snow sheep. In Tanzania 10,000 to 25,000 elephants are killed every year for the ivory tusks. President Obama has already suggested the use of drones to the Tanzania Ambassador. In July 2013, President Obama issued an executive order to provide $10 million dollars to troubled countries to combat the poaching problem in Africa. If you enjoy safari's it's time to write a proposal for funding. One person I know has obtained investor funding to purchase several of the long endurance Penguin drones with an integrated Cloud Cap EO/IR sensors. He purchased the Penguin from the UAV Factory in Latvia and US based Cloud Cap did the sensor integration. A UAV equipped with an EO/IR sensor would alow for day and night operations to discover illegal poachers.

The International Whaling Commission (IWC) banned commercial whaling in 1986. Most blubber can be used for human consumption and the other parts can be used for pet food. Japan uses drones to photograph Japanese whaling ships and turn the poachers into the authorities.

In my research I discovered 4 animal rights groups that are using UAVs to Monitor animal activities and report violations to authorities. The World Wide Fund is in Africa and put a request for proposal out for UAVs. They monitor rhinos, elephants and tiger poaching. The People for Ethical Treatment of Animals (PETA) in the USA combats those who gun down deer and doves. They film illegal hunting activity and turn it over to law enforcement. Animal Liberation Australia flies over farms to monitor cruelty to farm animals. Finally, there is Showing Animals Respect and Kindness (SHARK) who had their drone shot down while monitoring a pigeon hunt in South Carolina.

Drone Mapping Biz

Payload = Optical

A site survey is an inspection of an area where work is to be performed. Typically they are used in the construction industry but can also be used for drilling, farming and siting cell phone towers. Engineers use topographic maps to figure out how to move the earth around and architects use them to plan how the finished project will look. Features such as the pathways, roads, wells, the electricity supply cables, beaches, rocky outcrops and vegetation appear on a topographic map. The first step in a site survey is to obtain a topographic map which is a very detailed representation of elevations usually using contour lines to connect points with equal heights. Features such as the village, pathways, roads, wells, the electricity supply cables, beaches, rocky outcrops and vegetation appear on a topographic map. A topographic map is normally constructed using personnel to take measurements of building dimensions, take pictures and make notes and observations. They look for obstacles that may cause problems during construction. They position stakes at various points and make distance, angle, slope and height measurements using theodites, levels, tape measures, a compass and ranging rods.

A similar map can be obtained using photogrammetry where a set of pictures is taken from different angles and processed to formulate a digital elevation model (DEM) image. Dronemapper.com offers this as a service where you can upload your photos and receive a compiled image. Dronemapper has a detailed explanation about the format, elevation, geo-tagging, etc. MosaicMill makes products called EnsoMOSAIC UAV and EnsoMOSAIC 3D. EnsoMOSAIC is photogrammetry software that does not stitch images together. Stitched images are visually pleasing but can contain distortions. This software does not stich the images. Instead it calculates an elevation model and renders the image. EnsoMOSAIC 3D is digital photogrammetric software that processes photos into a stereoscopic 3D image.

There are a few open source products called Mapmill and MapKnitter. Instead of sorting through thousands of images for your map, Mapmill will pick out the best ones. Then you can upload these images into MapKnitter which will figure out where the images can be placed on a map. Another series of products I found is called Meshlab and Bundler. The first step is to install Meshlab (open source), a 3D mesh processing software system. Next you install Bundler, which orientates the images. Complete examples and a tutorial will help you get started.

Figure 49 SmartOne Personal Aerial Mapping System.
Courtesy of Troy Built Models.

You can buy a drone with an optical sensor from Aeromapper. This small drone has an option for a 24 megapixel camera with a resolution of 2 cm/pixel, or a Tetracam NIR sensor and covers up to 10 km^2 in 40 minutes and automatically snaps pictures. Prices range from $2700 to $9990. Custom payloads can also be installed but you will need to call for a quote. You will need to do your own processing of the data to stitch it together. Another ready to go platform is the SmartPlanes Personal Aerial Mapping System (PAMS). They make a drone called SmartOne that can produce photomaps or even 3D models. The photos are stitched together to form a mosaic. The drone can cover 250-300 acres in a single flight and 2400 acres a day.

A company called Precision Hawk does it all for you. They can integrate any type of sensor and allow you to program the survey area from home. Then you go to the field, start the engine, hand throw into the air and the drone automatically flies the programmed pattern. When finished, the drone comes back to the point of origin and lands itself on its belly. Take the drone home and you can upload the data to the company. They will stitch together the images and send you the finished product. The novel thing about this package is that you do not require any pilot training. You simply start the engine and hand throw into the air. Another novelty is that the sensors are plug and play. They use "intelligent sensor specific flight planning" You simply input the boundaries, sensor, resolution and the flight plan software does the rest. The software will detect the sensor and trigger the optimal overlap.

Another company called Pteryx sells a drone with mapping software included. This drone takes multiple images and assembles them into a 3D image. This drone is made of composites and hard wood. Takeoff and landing is automatic so you don't even need flight instruction. Just select a pre-programmed mission and push the takeoff button. Pyteryx will fly the mission and return to you via parachute.

Other applications for mapping are counting to quantify economic activity. You can count the number of cars in the mall parking lot. Crowded parking lots could be an indicator of increased sales. The same goes for trains. According to former Treasury Secretary John Snow who was CEO of CSX railroad, a typical efficient train is about 100 – 110 cars.

Figure 50 Pyteryx Drone with Mapping Software.
Courtesy of Troy Built Models

During recessions trains have been observed to have 60-70 cars. So if the economy is doing well, a train will have 100 or more cars. This observation has been tested over a number of years an found to be a good indicator of economic activity. What about international economic activity. How about counting the number of ships and their frequency. And for container ships how high are the containers stacked 4, 5 or 6 high? When the boxes are stacked 7 high, carriers add another boat.

A company called DroneMetrix in Australia, where drones can be operated commercially, was hired by a copper-gold mine company to map the mines topology to make sure walls don't collapse and to monitor other safety issues. They are getting accuracies of up to 25 mm with their TopDrone-100. The drones also fly over the mine every month to measure remaining resources. A process that would normally take a number of people several days to complete can be accomplished in one hour. The company has a team of 9 employees that include pilots, mapping experts and surveyors. They opened up an office in Poland and are looking for other opportunism in Europe, Asia and the Middle East.

Drone Shark Watch Biz

Payload = Optical & Thermal

Every year about 100 shark attacks are reported worldwide. The countries with the largest number of attacks are the US and Australia, Australia has the highest number of fatalities with Western Australia being the deadliest place in the world. If you want to determine the most likely places for attacks you can examine the International Shark Attack File (ISAF) or the Global Shark Attack File (GSAF).

Figure 50 Galapagos Shark. Courtesy of NOAA

Sharks bite to explore and will usually swim away after one bite. Attacks are actually classified into three categories; hit and run, sneak attack and bump and bite. A hit and run attack is usually non-fatal and the shark bites and leaves. During a sneak attack, the victim does not see the shark and may sustain multiple bites. This is the most fatal type. A bump and bite attack occurs when the shark circles and bumps the victim before biting. Repeated biting occurs and can be severe or fatal.

There is currently no way to prevent a shark attack. So this is an area for an entrepreneur to use a drone to stand watch over the ocean for sharks and provide a real time warring to those in danger. Australia is already experimenting with this concept and has small drones patrolling Shark Bay. Several Australian Universities are conducting work on signal processing algorithms that will automatically detect and classify sharks. Another element that could be added to the architecture is an audible warning system with multiple languages to warn people in the water of a shark in the area. Maybe you tune your radio to a shark watch frequency and surfers wear wireless headsets to listen in or maybe the drone attacks the shark by dropping some kind of chemical repellent.

Manned helicopters have a difficult time spotting sharks 2.5 meters below the surface. Cameras with polarized lenses could see sharks 6-8 meters below the surfacee. Higher altitudes would give broader views of the search areas. Video feed could be sent to lifesavers to monitor the beaches in real time. If you want a complete unmanned solution, use an unmanned underwarter vehicle (UUV) with the drone. The UUV could patrol and see underwater and through a tethered feed to a floating antenna, send the video feed to the drone which caould relay the imagery to a lifesaver.

Drone Solar Panel Biz

Payload = Near Infrared and Thermal Infrared

The largest solar panel facility in the world is the Solar Energy Generating System (SEGS) plant in the California Mojave Desert. It generates an enormous 354 MW of electricity, enough to power 232,500 homes. In Kern and LA Counties another project is underway that will generate 570 MW and will power 400,000 homes. In the USA the top five states using solar power are California, Nevada, Hawaii, Utah and Idaho. Around the globe the top five are Germany, Italy, Japan, Spain and the US. Figure 51 shows the solar map for the USA.[48]

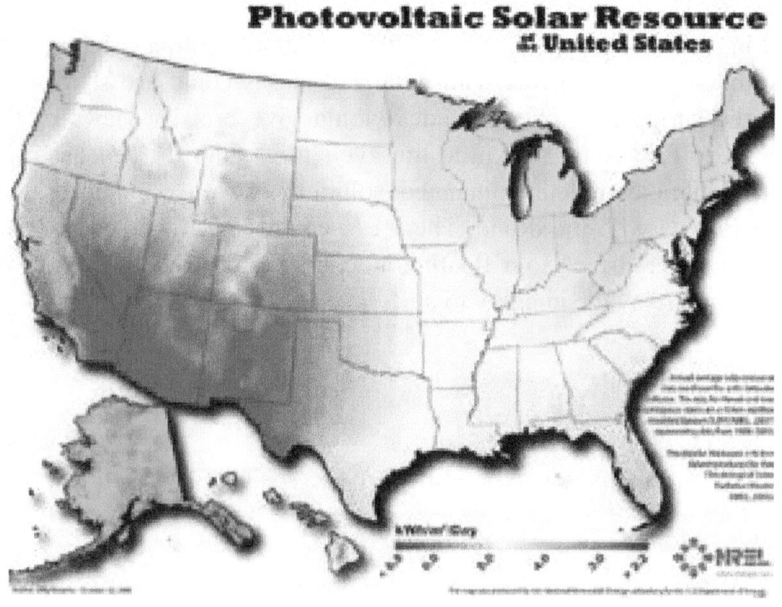

Figure 51 Solar Power Map of the USA

From this figure you can see that the best place to do your solar panel inspection business in the US is California, Arizona, Nevada and New Mexico. The capacity in Arizona is 283 MW and as of January, 2013, there were 19 projects seeking permission to add 13.45 GW. In 2012 solar energy installations increased by 80% gaining 3.3 gigawatts of capacity and 31 gigawatts of solar power was installed. Solar power is growing at an exponential rate so this is a very attractive field to gain new customers in the US and around the world.

The concept of operation of a solar panel is to use a photovoltaic cell to convert sunlight directly into electricity. The photovoltaic effect is a process by which light from the sun hits a solar cell and is absorbed by a semiconducting material such as crystalline silicon. The photons in the sunlight knock electrons loose from their atoms, allowing them to flow freely through the material to produce direct electric current (DC) electricity. For household or utility use, an inverter must be used to convert the electricity to alternating current (AC).[49] Solar panel construction consist of arranging a number of solar cells into an array. The arrays are coated with glass or another laminate to protect the cells from damage. Utility companies require thousands of panels. The largest solar panel facility in America is in Boulder City, NV. A picture of President Obama giving a speech at this plant is shown in Figure 52

Over time anomalies occur such as interconnection problems, defective bypass diodes, short circuits or cracks. These anomalies will show up as hot spots using a thermal IR sensor. The thermal image should be taken under various loaded and unloaded conditions. Once an anomaly is detected, the panel should be inspected visually and tested electrically to confirm the defect. If the image is GPS tagged (geo-taged) this can help localize the damaged cells or panels over large areas. One advantage to this inspection technique is that the thermal panels do not have to be shut down for inspection. Example thermal imagers can be purchased from FLIR (Tau), UAV Vision (CM Series) payloads and the IAI (Micro Pop).

Figure 52 President Obama Speech a Boulder City, NV Plant

Imagery can be directly interpreted so no processing is required. These imagers are very small (in the hundreds of grams) and can easily be installed on a small drone.

When you start your drone solar panel inspection business you will have to get up to speed on image interpretation. You should take a course on thermography to make sure you are interpreting the images correctly and know when the imager is not working correctly. After training, start with something familiar and get used to your equipment. You will most likely want to have the imagery data linked down to a laptop in real time so you can record, analyze. and determine where you have discovered anomalies.

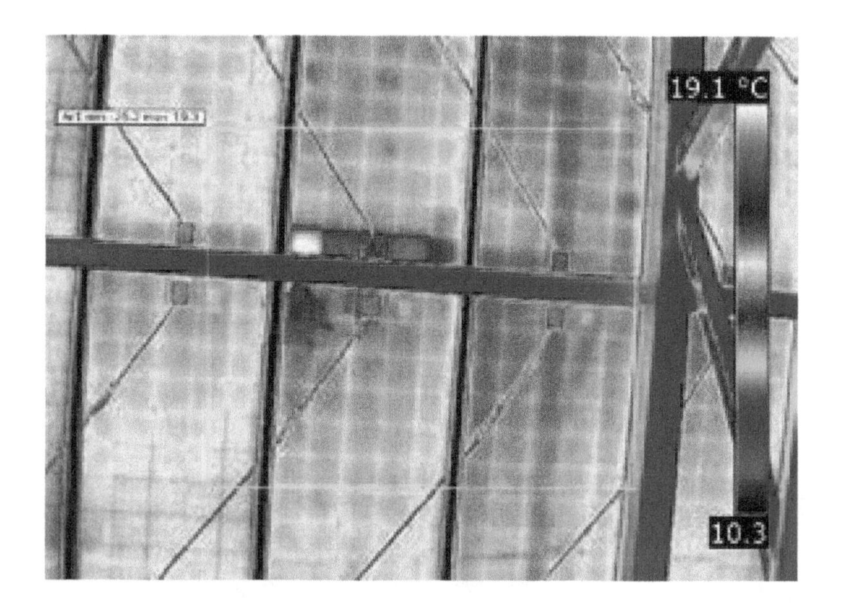

Figure 53 Defective Solar Panel.
Courtesy of FLIR

The best time to do inspections are after installation and periodic inspections during operation. Start with a small company, achieve some successes and keep moving up. Investigate projects that are close to starting up operations and offer your service. Determine your pricing by hour or by area. Offer your service for free when you start and after you have a solid handle on your reporting start charging.

You should also consider using a Short Wave Infrared or a Near Infrared (NIR) camera for this business. SWIR can detect electroluminescence which is an emission of light in the SWIR band in response to the passage of current. Defective solar cells are clearly visible. Not only can you see the cell that is bad, you can see great detail inside each cell and detect anomalies like cracking very easily. With the enormous growth, this can be a very profitable business. Think about partnering with solar panel construction companies once you have a portfolio. Offer your service after initial installation then meet with the customer to negotiate periodic inspections.

Drone Cargo Delivery Biz

Payload = Cargo

Fred Smith, the founder of FedEx wants to get into the drone cargo business as soon as possible. In a 2009 blog post he said:

Figure 54 Drone Cargo Delivery Concept

"Unmanned cargo freighters have lots of advantages for FedEx: safer, cheaper, and much larger capacity. The ideal form is the 'blended wing.' That design doesn't make a clear a distinction between wings and body, so almost all the interior of both can be used for cargo. The result is that the price premium for air over sea would fall from 10X to 2X (with all the speed advantages of air)."

The post goes on to say: "the key thing is having no people on board, not even as backup. A single person in the craft requires a completely different design, along with radically different economics and logistics. The efficiencies come with 100% robotic operation."

In my opinion, it could happen in the next 5 to 10 years for sure. Most likely we will only have two airports that will accommodate these unmanned cargo aircraft. One on the East and one on the West coast. They will not mix it up with too much traffic as the concept needs to prove itself out before it is accepted. I believe it is not a matter of if but a matter of when unmanned cargo aircraft will be moving freight around the world.

Figure 55 K-Max Unmanned Helicopter. Courtesy of US Navy

The K-Max unmanned helicopter has already made a hit as a cargo carrier. This helicopter can carry 6,000 pounds of payload, more than its weight and more than any drone in the world. It was designed for hazardous missions and can be used to deliver supplies or hoist up vehicles and move them around. It has already delivered 2 million pounds of cargo during 600 unmanned flights with over 700 hours of operation. It is 50 ft. long, 13 ft. high, has a maximum takeoff weight of 12,000 pounds and cruises at 80 knots. The K-Max has a ceiling of 11,500 ft., more than 2,000 ft. higher than the average mid-sized helicopter.

Drone Instant Gratification Biz

Payload = Cargo

You order a book from Amazon, the order is filled and within an hour or two a drone arrives with your book. CEO Jeff Bezos says "the holy grail of shiping - same day delivery - is tantlizingly within reach." Amazon already offers same day delivery in select cities. For about $10-12 you can have a book delivered in 9 hours. Drones will take a giant leap ahead of same day delivery by delivering in a few hours. An ideal application is DVD movie rentals. Imagine ordering a very light payload Netflix movie and receiving it within a few hours. You will need to keep the products light as payload is a premium on drones. They are being made with lighter composite materials so payload limits can be increased. A fixed wing drone will require parachutes and multicopters can actually descend to the customer to take possession. In July 2013, a dry cleaning company in Philadelphia demonstrated the delivery of a shirt that was dry cleaned and delivered by a drone.

Instant gratification will open new markets that will be compatible with existing business. Think about document delivery and the ways that you can increase rates above what you already have for faster delivery methods. If you don't have an existing business you can start one based on the instant gratification concept. Find a company that is suitable for this concept, offer your services, do a demonstration, record your work to build a portfolio and you will be off and running.

Walmart is hinting at a strategy that uses its customers to act as couriers for same day delivery. Walmart will offer a discount for customers to do same day delivery. Drones will leap ahead of all of this by offering delivery within hours. Documents, letters and even pizzas delivered in a short time interval will cause a dramatic change.

Drone Avalanche Biz

Payload = Microwave Transceiver

If you like skiing and want to run a business then this is the one for you. An avalanche is a rapid flow of snow down a slope. Over the last 10 winters in the United States an average of 25 people died in avalanches every year. Every fatal accident is investigated and reported to the Colorado Avalanche Information Center (CAIC) in Boulder, CO.

Figure 56 Snow Avalanche in Switzerland
Courtesy of Swiss Federal Institute for Snow and Avalanche Research

There is no way to determine the number of people caught or buried in avalanches because non-fatal avalanches are under reported. The average yearly property loss is $31,200.[50] If you visit the CAIC website you can see exactly where each has occurred. A total of 34 were killed in 2012 in all states with the greatest number of avalanches occurring in Colorado where 14 were killed in 2012. Other states with 2 or more deaths due to avalanches are Washington State, Wyoming and Utah.

Loose snow accounts for only a small percentage of avalanches. Slab avalanches (cohesive plates) are the most lethal. Each year avalanches kill 150 people worldwide. In 90% of avalanche accidents, the victim or someone in the victims party causes a slide. The human body is three times denser than avalanche debris and will sink quickly. Noise does not trigger avalanches, they are caused by four factors; steep sope, snow cover, weak layer and a trigger. Avalanches can reach speeds of 80 mph in 5 seconds. An avalanche is often triggered when a persons body weight provides just enough extra stress to collapse the weaker layer below. Avalanche risk is greatest 24 hours following a snowfall of 12 inches or more. That is when you want your drone business should be placed on alert for your customer. If a victim can be rescued within 18 minutes, the survival rate is greater than 91%. The survival rate drops to 34% in burials between 19 and 35 minutes. The primary cause of death is asphyxiation (severe lack of oxygen).[51]

Some prestigious ski resorts have helicopters on standby with an average operating cost of 2000 Euros per hour. A low cost solution would be for you to start a drone rescue business and offer your services to ski resorts. One company based in France is already doing this. Delta Drones has launched a mountain avalanche protection drone. The drones have sensors to ascertain terrain type, snow depth and layers and a transceiver for avalanche victim recovery. The drone is capable of targeted avalanche triggering and has a loudspeaker that can speak French, Russian and English. As of July 2013 the drone is being tested in Isere, France.

Drone Asset Protection Biz

Payload = Optical & Infrared

If you already own or want to start a security company this is the one for you. I was just contacted by a security company in a Caribbean country. They want to add a drone to monitor certain assets. The concept was to send a drone around to periodically monitor for any abnormal activity. I asked for their endurance requirements and the answer was 90 minutes. A multicopter is more suitable for this application as you may want to hover and stare at an area where you detect activity. Most multicopters are battery powered and have much less than one hour endurance. So I came back with a recommendation for the Rotomotion SR 30 which is a gasoline powered internal combustion engine helicopter with an endurance of 1.5 to 3 hours.

A company called Drone America already offers security services and will customize a drone with cameras, sensors and specific payloads. The client specifies the equipment and Drone America does the specialized engineering to perform the integration. They partner with other companies and offer a design review, design change order and final integration. Their drones have programmable autopilots so you could easily specify the flight path for the drone to search for activity.

This business can be an easy start. Go online and order a security uniform with your business name on it and get an official looking hat. Don't forget the security badge. Add a custom design with a picture of a drone and start to make calls on customers with high end homes and commercial companies that close down on the weekends. Visit clients that already have security guards and explain drone advantages over a security camera (limited field of view, incomplete area coverage, etc.).

Drone Medical Supply Biz

Payload = Medical Supplies

A Silicon Valley startup company called Matternet is creating a network of drones to deliver food and medical supplies to remote areas with no roads in Africa. There are over one billion people in the world without access to all-season roads. Matternets' vision is to create a new paradigm (do you really need roads?) for transportation. Using small battery powered drones that have less than one hour of flight time, the concept is to have drones travel between relay stations where the batteries can be swapped out for fully charged ones. Once they arrive at the station, the battery is swapped out until it reaches its final destination.

Figure 57 Gas and Food Delivery via Parachute
Courtesy of the USAF

Instead of a single drone with multiple stations, the company wants to create a network of multiple drones to expand the coverage area. Matternet is already carrying a 6 lb payload in Haiti and the Dominican Republic. Matternet has already partnered with an organization to help manage a drone fleet to deliver medical supplies in West Africa.

In June 2013, India was hit by the worst flooding in more than 50 years. The death toll could reach 1000 and more likely 10,000. There were 5100 still missing at the writing of this book and no final statistics were available. My research revealed ten startup companies in India that are delivering food and medical supplies to victims in isolated areas using multicopter drones.

Obviously your challenge here is to get permission to deliver medicine. Maybe you need a Medical Doctor or Pharmacologist on your staff that can fill prescriptions. Will you be for profit or non-profit. There is certainly a market to do real time delivery of prescriptions to those that can't walk or drive and have to order through the mail.

When someone gets in medical trouble a drone could fly to the victim and give the doctor a birds eye view. Call 911 and the drone takes off to the precise location of the victim. The doctor could remotely evaluate the patient and talk to people around or give instructions using microphones and speakers.

A company in Germany called Definetz is launching a service to deliver defibrillators to heart attack victims. Their deficopter can fly at 43 mph, go 6 miles and drop off a defibrillator. The cost is $26,000. And finally, Bill Gates has pledged $10 billion to vaccine delivery to remote areas. $100,000 is being provided to 17 different initiatives for a total of $1.7 million. Harvard-MIT Health Sciences and Technology won one of the grants to use drones to deliver vaccines.

Drone Boating Biz

Payload = Optical & Near Infrared

If you are a boater, how many times have you operated in the fog and wondered if you were going to hit anything. Well it's time to bring out your drone to help see through the weather. We spoke about thermal cameras in previous business applications. Thermal cameras are good at seeing heat emissions and are severely limited by fog. A Near Infrared (NIR) camera can actually see through the fog. You have to make sure you have a near IR (NIR) or shortwave IR (SWIR) camera that operates in the 0.7 to 1.2 micrometer range. A NIR camera operates by detecting reflected light. Longer wavelengths like medium and far IR thermal sensors do not detect reflected light because the longer the wavelength, the less molecules and small particles in the air there are to scatter the light. In Figure 58 you can see the comparison between optical and SWIR on a foggy day over the water. A multicopter is more suitable for this application as it can slow down and even stop (hover) with the slow velocities of a boat. Consider selling a multicopter equipped with a SWIR to boaters.

Figure 58 Optical and Short Wave Infrared Comparison
Courtesy of NASA

Drone Oil Exploration Biz

Payload = Optical, Magnetomter, Laser, Infrared

Oil exploration involves integration of information from many fields including geology. geochemistry, seismology, electromagnetism and others. There are several types of surveys that can be taken and when integrated together indicate the presence of oil. Geomagnetic surveys using magnetometers record changes in the earth's magnetic field caused by different types of rocks with different magnetic pulls. Gravity surveys with a gravimeter help determine where porous rocks can be found. Seismic surveys transmit sound that is recorded by a hydrophone and can record underground rock formations. If all of these survey types are compiled and analyzed, indications of an oil or gas reserve may be determined. When the drilling commences, samples of rocks are brought to the surface and analyzed to measure their magnetic, electrical and radioactive properties.

Drones have already carried out geomagnetic and gravity surveys. In Norway, geologists are using drones for oil exploration with other sensors. The drones use laser scanners, IR sensors and digital cameras to build 3D models of the terrain. The rocks and minerals on the surface can reflect what is beneath the ground.

Costs are the most significant advantage of drones. The cost of drone operations are at least one tenth of the cost of operating a manned aircraft. If you want to get into this business you need to learn about geomagnetic and gravity surveys, equip your drone with the proper sensors and practice on areas that already have the surveys accomplished. Compare your results to theirs and when you are ready, start to offer your services to oil and gas exploration companies. Your major advantage will be reduction in cost. This is the most important element to your business case. Develop a portfolio of surveys for your next customer.

References

[1]FAA Fact Sheet, Unmanned Aircraft Systems (UAS), July 2011

[2]DoD, "Principal Wars in Which the United States Participated: U.S. Military Personnel Serving and Casualties"; and Operation Iraqi Freedom—U.S. Casualty Status (www. defenselink.mil, accessed Oct. 31, 2004.)

[3]Department of Defense and Veterans Administration.

[4]FAA Website, Home>News>News & Updates, July 26, 2013

[5]National Telecommunications & Information Administration (NTIA), U.S Department of Commerce.

[6]U.S. Census Bureau, Economics & Statistics Administration, May 2013.

[7]Measuring Broadband America, February 2103, A Report on Consumer Wireline Broadband Performance in the U.S., FCC's Office of Engineering and Technology and Consumer and Governmental Affairs Bureau.

[8]The Impact of Broadband Speed and Price on Small Business, Columbia Telecommunications Corporation for the Small Business Administration Office of Advocacy, Contract No. SBAHQ-09-C-0050, November 2010.

[9]U.S. Department of Energy, Energy Efficiency and Renewable Energy. Office of Transportation and Air Quality, U.S. Environmental Protection Agency.

[10]A. Noth, R. Siegwart, Design of Solar Powered Aircraft for Continuous Flight, Swiss Federal Institute of Technology, Zurich, Version 1.0, 2006

[11]H. Tennekes, The Simple Science of Flight, From Insects to Jumbo Jets, MIT Press, Cambridge, Mass. London England, 1996.

[12]R. Jones, H. Ougham, H. Thomas, S. Waaland, The Molecular Life of Plants. John Wiley and Sons, 2013

[13]Federation of American Scientists, Tutorial on Vegetation Applications.

References (cont.)

[14]E. Flynn, Using NDVI as a Pasture Management Tool, University of Kentucky Masters Thesis, Paper 412, 2006

[15]U.S. Department of Agriculture

[16]Ibid

[17]W. Simonson, H. Alen, D. Coomes, Detection of Fungal Infection of Plants by Laser-Induced Fluorescence: An attempt to Use Remote Sensing, Forest Ecology and Con-servation Group, Department of Plant Sciences, University of Cambridge, Cambridge, UK. Oct 26, 2012.

[18]Michigan Department of Community Health, grant number 5 U01 OH007306 from the U.S. Centers for Disease Control & Prevention, National Institute for Occupational Safety and Health (CDC-NIOSH).

[19]S. Ellis, M. Boehm, Plants Get Sick Too, Plants Get Sick To, An Introduction to Plant Disease, Ohio State University, Department of Plant Pathology.

[20]Wheat: Grappling With Ingrained Problems," Agricultural Research, Vol. 74, No. 7, p. 16, July, 1990.

[21]Townsend, G.R., Diseases of Beans in Southern Florida, University of Florida Agricultural Experiment Station, Bulletin 336, September, 1939.

[22]Ivanoff, S.S., "Spinach and Onion Diseases in the Winter Garden Region of Texas," Plant Disease Reporter, Vol. 21, pp. 114-115, 1937

[23]Rinehold, John, and J.J. Jenkins, Oregon Pesticide Use Estimates for Seed Crops and Specialty Crops, 1992, Oregon Pesticide Impact Assessment Program, Oregon State University, 1993

[24]U.S. Department of Agriculture, Agricultural Research Service.

[25]M. Sheridan, Catholic Research Services, April 16, 2013

References (cont.)

[26]A. Mahlein, U. Steiner, C. Hillnhutter, H. Dehne, E. Oerke, Hyperspectral Imaging for Ssmall-scale Analysis of Symptoms Caused by Different Sugar Beet Diseases, Springer Link, January 2012

[27]G. Menz, W. Kuhbauch, M. Braun, J. Franke, Spatiotemporal Dynamics of Stress Factors in Wheat Analyzed by Mutisensoral Remote Sensing and Geo Statistics, DFG Research Training Group, University of Bon, Germany.

[28]S. Sankaran, J. Maja, S. Buchanon, R. Ehsani, Huanglongbing (Citrus Greening) Detection Using Visible, Near Infrared and Thermal Imaging Techniques, Sensors Basel) 2013, 2117-2130, February 6, 2013

[29]US Department of Agriculture.

[30]University of Florida, IFAS Extension, Publication #CIR1256

[31]L. Williams, Irrigation of Winegrapes in California, Winery and Vineyard Journal, November 2001.

[32]FAA UAS Listening Session, FAA 04.01.13, April 3, 2013.|

[33]Iowa Department of Transportation's Office of Aviation.

[34]Everitt, J. H., D. E. Escobar, M. A. Alaniz, M. R. Davis, and J. V. Richerson1996. Using spatial information technologies to map Chinese tamarisk (Tamarix chinensis) infestations. Weed Sci. 44:194–201.

[35]Mirror News, July 19, 2013.

[36]The European Association for the Conservation of the Geological Heritage, ProGEO, 2010

[37]Wilson, D. R. (1982), Air photo interpretation for Archaeologists, (2nd ed.)

References (cont.)

[38]National Interagency Fire Center

[39]On Scene Coodinator Report, Deepwater Horizon Oil Spill, Sept 2011 [40]Department of the Interior, National Incident Command, Interagency Solutions Group, Flow Rate Technical Group, March 10, 2011

[41]US Department of Energy, Energy Efficiency and Renewable Energy, Wind Program, January 17, 2013

[42]Global Wind Energy Council

[43]Caithness Windfarm Info Forum, www.caithnesswindfarms.co.uk

[44]Lux Research

[45]L. Cheng, G. Tian, Comparison of Nondestructive Testing Methods on Detection of Delaminations in Composites, Journal of Sensors, Volume 2012, Article ID 408437.

[46]A. Keemink, MSc Report, University of Twente, EEMCS/Electrical Engineering Control Engineering, January 2012

[47]American Society for Testing and Materials (ASTM) D4580-86 Standard Practice for Measuring Delaminations in Concrete Bridge Decks by sounding.

[48]US Depatment of Energy

[49]Ibid.

[50]Colorado Avalanche Information Center

[51]USGS, FEMA and Red Cross

Appendix 1 FAA Advisory Circular 91-57

Subject: MODEL AIRCRAFT OPERATING STANDARDS

1. PURPOSE. This advisory circular outlines, and encourages voluntary compliance with, safety standards for model aircraft operators.

2. BACKGROUND. Modelers, generally, are concerned about safety and do exercise good judgment when flying model aircraft. However, model aircraft can at times pose a hazard to full-scale aircraft in flight and to persons and property on the surface. Compliance with the following standards will help reduce the potential for that hazard and create a good neighbor environment with affected communities and airspace users.

3 OPERATING STANDARDS.

a. Select an operating site that is of sufficient distance from populated areas. The selected site should be away from noise sensitive areas such as parks, schools, hospitals, churches, etc.

b. Do not operate model aircraft in the presence of spectators until the aircraft is successfully flight tested and proven airworthy.

c. Do not fly model aircraft higher than 400 feet above the surface. When flying aircraft within 3 miles of an airport, notify the airport operator, or when an air traffic facility is located at the airport, notify the control tower, or flight service station.

d. Give right of way to, and avoid flying in the proximity of, full-scale aircraft. Use observers to help if possible.

Appendix 2 14 CFR 21.95 Experimental Certificates

Aircraft to be used for market surveys, sales demonstrations, and customer crew training.

(a) A manufacturer of aircraft manufactured within the United States may apply for an experimental certificate for an aircraft that is to be used for market surveys, sales demonstrations, or customer crew training.

(b) A manufacturer of aircraft engines who has altered a type certificated aircraft by installing different engines, manufactured by him within the United States, may apply for an experimental certificate for that aircraft to be used for market surveys, sales demonstrations, or customer crew training, if the basic aircraft, before alteration, was type certificated in the normal, acrobatic, commuter, or transport category.

(c) A person who has altered the design of a type certificated aircraft may apply for an experimental certificate for the altered aircraft to be used for market surveys, sales demonstrations, or customer crew training if the basic aircraft, before alteration, was type certificated in the normal, utility, acrobatic, or transport category.

(d) An applicant for an experimental certificate under this section is entitled to that certificate if, in addition to meeting the requirements of § 21.193 —

(1) He has established an inspection and maintenance program for the continued airworthiness of the aircraft; and

(2) The applicant shows that the aircraft has been flown for at least 50 hours, or for at least 5 hours if it is a type certificated aircraft which has been modified. The FAA may reduce these operational requirements if the applicant provides adequate justification.

Appendix 3 FAR Part 91.319 Experimental Limits

(a) No person may operate an aircraft that has an experimental certificate--
(1) For other than the purpose for which the certificate was issued; or
(2) Carrying persons or property for compensation or hire.

(b) No person may operate an aircraft that has an experimental certificate outside of an area assigned by the Administrator until it is shown that--
(1) The aircraft is controllable throughout its normal range of speeds and throughout all the maneuvers to be executed; and
(2) The aircraft has no hazardous operating characteristics or design features.

(c) Unless otherwise authorized by the Administrator in special operating limitations, no person may operate an aircraft that has an experimental certificate over a densely populated area or in a congested airway. The Administrator may issue special operating limitations for particular aircraft to permit takeoffs and landings to be conducted over a densely populated area or in a congested airway, in accordance with terms and conditions specified in the authorization in the interest of safety in air commerce.

(d) Each person operating an aircraft that has an experimental certificate shall--
(1) Advise each person carried of the experimental nature of the aircraft;
(2) Operate under VFR, day only, unless otherwise specifically authorized by the Administrator; and
(3) Notify the control tower of the experimental nature of the aircraft when operating the aircraft into or out of airports with operating control towers.

(e) No person may operate an aircraft that is issued an experimental certificate under §21.191 (i) of this chapter for compensation or hire, except a person may operate an aircraft issued an experimental certificate under §21.191 (i)(1) for compensation or hire to-
(1) Tow a glider that is a light-sport aircraft or unpowered ultralight vehicle in accordance with §91.309; or (2) Conduct flight training in an aircraft which that person provides prior to January 31, 2010.

(f) No person may lease an aircraft that is issued an experimental certificate under §21.191 (i) of this chapter, except in accordance with paragraph (e)(1) of this section.

(g) No person may operate an aircraft issued an experimental certificate under §21.191 (i)(1) of this chapter to tow a glider that is a light-sport aircraft or un-powered ultralight vehicle for compensation or hire or to conduct flight training for compensation or hire in an aircraft which that persons provides unless within the preceding 100 hours of time in service the aircraft has-
(1) Been inspected by a certificated repairman (light-sport aircraft) with a maintenance rating, an appropriately rated mechanic, or an appropriately rated repair station in accordance with inspection procedures developed by the aircraft manufacturer or a person acceptable to the FAA; or
(2) Received an inspection for the issuance of an airworthiness certificate in accordance with part 21 of this chapter.

(h) The FAA may issue deviation authority providing relief from the provisions of paragraph (a) of this section for the purpose of conducting flight training. The FAA will issue this deviation authority as a letter of deviation authority.
(1) The FAA may cancel or amend a letter of deviation authority at any time.
(2) An applicant must submit a request for deviation authority to the FAA at least 60 days before the date of intended operations. A request for deviation authority must contain a complete description of the proposed operation and justification that establishes a level of safety equivalent to that provided under the regulations for the deviation requested.

(i) The Administrator may prescribe additional limitations that the Administrator considers necessary, including limitations on the persons that may be carried in the aircraft.

Appendix 4 CFR Sec 21.25 Type Certificate

Issue of Type Certificate: Restricted Category Aircraft.

[(a) An applicant is entitled to a type certificate for an aircraft in the restricted category for special purpose operations if he shows compliance with the applicable noise requirements of Part 36 of this chapter, and if he shows that no feature or characteristic of the aircraft makes it unsafe when it is operated under the limitations prescribed for its intended use, and that the aircraft--]
(1) Meets the airworthiness requirements of an aircraft category except those requirements that the [FAA] finds inappropriate for the special purpose for which the aircraft is to be used; or
(2) Is of a type that has been manufactured in accordance with the requirements of and accepted for use by, an Armed Force of the United States and has been later modified for a special purpose.
(b) For the purposes of this section, "special purpose operations" includes--
(1) Agricultural (spraying, dusting, and seeding, and livestock and predatory animal control);
(2) Forest and wildlife conservation;
(3) Aerial surveying (photography, mapping, and oil and mineral exploration);
(4) Patrolling (pipelines, power lines, and canals);
(5) Weather control (cloud seeding);
(6) Aerial advertising (skywriting, banner towing, airborne signs and public address systems); and
(7) Any other operation specified by the [FAA].

Appendix 5 Scan Eagle Type Certification

August 5, 2013

This data sheet, which is part of Type Certificate No. Q00017LA, prescribes conditions and limitations under which the product for which the type certificate was issued meets the airworthiness requirements of the 14 Code of Federal Aviation Regulations (14 CFR).

Type Certificate Holder Insitu Inc. 118 East Columbia River Way Bingen, WA 98605 USA

I. Model ScanEagle X200 (Restricted Category UAS) Approved July 19, 2013 (See NOTES Section)

UAS This is an Unmanned Aircraft System (UAS) that is comprised of the air vehicle and the transportable ground control station. Unmanned Aircraft Dimensions Wingspan 10.2 Ft (3.11 m) Length 4.5 Ft (1.37 m)

Engine (1) Northwest UAV, Block D Hush FAA Engine Type Certificate: None Engine type: Normally-aspirated, carbureted, two-stroke, direct drive, air cooled, single cylinder engine.

Fuel High Octane C-10 See FAA Approved Airplane Flight Manual (AFM) for additional information on approved fuel

Oil 2-cycle oil See FAA Approved AFM for additional information on approved oil

Fuel-Oil Mix Capacity 2.5 gallons

Engine Limits Max Takeoff Power 1.75 HP at 8500 RPM Max Continuous Power 1.75 HP at 8500 RPM Max Cylinder Head Temperature (CHT) 180°C Min CHT for flight 50°C Propeller and Propeller Limits. (1) APC, Propeller Model LP315135 FAA Propeller Type Certificate: None Propeller Type: 3-blade, chopped fiberglass and resin, 15 x 13 fixed pitch pusher Diameter (Nominal): 15 inches (38 cm) Pre-flight Static rpm requirement: Engine must successfully achieve the I-MUSE checklist item: Engine Performance Checked

Electric Generator 20 Volts, nominal 6.0 Amps, maximum 95 Watts, maximum

Backup Battery 19.2 Volts, nominal 1100 mA Hrs. Airspeed Limits (CAS) VNE 98 KTAS (181 km/hr) VNO 85 KTAS (157 km/hr) VA (Maneuvering Speed) Landing Speed (Closing Speed) – minimum Landing Speed (Closing Speed) – maximum 85 KTAS (157 km/hr) 25 KT (46 km/hr) 52 KT (96 km/hr)

Center of Gravity (C.G.) Range Center of gravity from datum Minimum: -1.97 in (-50 mm) at any weight Center of gravity from datum Maximum: -2.76 in (-70 mm) at any weight Reference Datum Location: 8.07 in (205 mm) forward from the aft edge of the fuselage module

Empty Weight C.G. Range None

Datum Located on centerline of airplane at wing trailing edge intersection: positive in the x direction to the nose; positive in the y direction to the right wing; and positive in the z through belly of the aircraft. (Insitu ScanEagle Datum Reference Drawing, 15 Jul 2013)

Mean Aerodynamic 9.5 in (241.3 mm) long with leading edge: Chord (MAC) x = -0.55 in (14.0 mm) from datum y = 27.8 in (706.7 mm) from datum.

Leveling Means When level—in the aircraft stand—and not moving, accelerometers should read: X-axis: 0.0 G+/-0.03 G Z-axis: -1.0 G+/-0.05 G

Maximum Weights Ramp 44 lbs (19.96 kg) Takeoff 44 lbs (19.96 kg) Landing Weight 44 lbs (19.96 kg)

Empty Weight 29.8 lbs. (13.5 kg)

Data UP-Link Frequencies 1.3 GHz, Commands for: aircraft control, sensor control, video control

Data DOWN-Link Frequencies 1.3 GHz, Report status on: aircraft, sensors, and video

Video Down-Link Frequency 2.4 GHz

NOTE: FCC license is required to utilize the above frequencies.

Computer Software I-MUSE Software Version 5.6.13

Minimum Crew (1) UAS pilot at the Ground Control Station (2) Personnel for launch and recovery Number of Seats (0) Not Applicable

Fuel Capacity Fuel System: 12.3 lbs. of fuel (5.6 kg) Unusable Fuel: 0.2 lbs. of fuel (0.1 kg) NOTE: Fuel capacity includes the oil mixed (50:1) with the fuel Oil Capacity Not Applicable Max. Operating Altitude 2000 ft. AGL (610 M) Control Surface Movements Outboard Elevon Up 30° Down 30° Deflections are +/- 2 degree Inboard Elevon Up 30° Down 30° Rudder Left 25° Right 25° Flight Endurance 18.5 Hrs.

Flight Limitations 1. Day Visual Flight Rules (VFR) in visual meteorological conditions (VMC) 2. Flight through visible moisture: PROHIBITED 3. Flight operations in icing conditions at assigned operational altitudes: PROHIBITED 4. Flight Pitch Attitude: +/- 45° 5. Flight Bank Angle: +/- 44° 6. Ambient Outside Air Temperature (OAT) a. Maximum OAT: 120°F / 49°C b. Minimum OAT at Altitude: -4°F / -20°C 7. Wind. See Note 5 8. Flight Operations. See Note 4 9. For this operation only one ScanEagle can be airborne at any given time 10. Over water operation: PERMITTED 11. Over land operation: PROHIBITED 12. An authorization for the specific location of operation issued by the Administrator is required and must be available at the control station. AFM number FAA-01-AFM, dated July 16, 2013 or later FAA approved revision, and certificate of airworthiness (C of A) must be available at the control station (reference FAA Memorandum, "Certification of Unmanned Aircraft", from AAL-7 to ANM-100L, dated June 19, 2013). Additionally, any certificates of authorizations or waivers must be available at the control station. 13. Only for operation in the designated Arctic Area as defined by the FAA Modernization and Reform Act of 2012, Section 332(d)(1). 14. Operation with inoperative instruments and equipment: PROHIBITED

Serial Numbers Approved 11-1313, 11-1453, 11-1458, 11-1459

Certification Basis Restricted Category Only 14 CFR part 21.25(a) (2) for the special purpose of aerial survey, 14 CFR part 36, amendment 29, Appendix G

Production Basis None

UAS Support Equipment: Launcher: Insitu P/N 090-000200R00. See Note 6 and Note 10. Skyhook: Insitu P/N 900-200402-005. See Note 6 and Note 10.

NOTES:

NOTE 1 Current weight and balance data, loading information, and a list of equipment included in the empty weight must be provided for each UAS at the time of original certification.

NOTE 2 Placards Required: None

NOTE 3 This UAS must be maintained in accordance with Unmanned Aerial Systems Maintenance Handbook, Version 2.0, dated September 2007, Document Number 026-010019, or later FAA accepted revision.

NOTE 4 UAS shall be operated under 14 CFR part 91, operating requirements, as mitigated. Operations shall be conducted in accordance with a waiver of flight regulations applicable to the operation, including but not limited to 14 CFR § 91.113, issued by the Administrator and specific to the intended operation, including geographical limitations.

NOTE 5 Wind Limitations: Ship launch wind over deck conditions: (a) Wind over deck conditions shall be determined by shipboard wind measurement and indication system. (b) Max gusts for launch and recovery: 5 Kts (5.75 mph, 9.26 kph) (c) Launches (including gusts): 1. 10 Kts from +/- 30° relative to the launcher centerline. 2. 20 Kts from +/- 20° relative to the launcher centerline. 3. Launches with tailwinds: PROHIBITED. (d) Recoveries (including gusts): 1. Port recoveries: a. 20 Kts from 320° to 350° relative to the ship centerline. b. 30 Kts from 320° to 330° relative to the ship centerline. 2. Starboard recoveries: a. 30 Kts from 10° to 40° relative to ship centerline. 3. Recoveries with tailwinds: PROHIBITED. (e) Wind limitations during flight: 1. Max winds (sustained plus gusts): 40 KIAS 2. Max gust component (gusts are considered any wind variations above the measured sustained value): 10 Kts

NOTE 6 Personnel Keep Out Zones. Typical exclusion zones apply for Launch and Recovery (SkyHook) as described in AFM and UAS Operations Manual.

NOTE 7 This Type Certificate Data Sheet (TCDS) is the principal document for ScanEagle Operation. For any operational discrepancies among the TCDS, AFM, Insitu ScanEagle Ops. HDBK, etc., this TCDS takes precedence.

NOTE 8 Restricted category aircraft may not be operated in a foreign country without the express approval of that country.

NOTE 9 This aircraft has not been shown to meet the requirements of the applicable comprehensive and detailed airworthiness code as provided by Annex 8 of the Convention on International Civil Aviation. This aircraft meets 14 CFR § 21.25(a)(2).

NOTE 10 For this restricted category type certificate, the part numbers of the Launcher and Skyhook must be those listed under UAS Support Equipment of this Type Certificate Data Sheet.

NOTE 11 Operations shall be conducted by properly certificated airmen who have completed training, checking, currency, and recency of experience requirements as approved by the Administrator.

Appendix 6 PUMA Type Certification

August 5, 2013

TYPE CERTIFICATE DATA SHEET No. Q00018LA

This data sheet, which is part of Type Certificate No. Q00018LA, prescribes conditions and limitations under which the product for which the type certificate was issued meets the airworthiness requirements of the 14 Code of Federal Aviation Regulations (14 CFR).

Type Certificate Holder AeroVironment, Inc. 181 W. Huntington Dr. Monrovia, CA 91016 USA

I. Model PUMA AE (Restricted Category UAS), Approved July 19, 2013 (See NOTES Section)

UAS This is an Unmanned Aircraft System (UAS) that is comprised of the air vehicle and the ground control station. Unmanned Aircraft Dimensions Wingspan = 9.2 Ft. Length = 4.7 Ft.

Engine (Propulsive Unit) (1) AeroVironment, Inc., Model 50333 (Electric) FAA Engine Type Certificate: None Propulsive Unit Type: 25 V, 13.5 Amp hour capacity, Lithium ion battery powered, direct drive electric motor Motor, Electric Sub-Assembly: Manufacturer: NeuMotor Model: 1910 1.0 HP Peak Power Direct Drive 10 oz. Wt. Motor, Controller Sub-Assembly: Manufacturer: Castle Creation Model: Phoenix HV-45 Type: Speed Controller 45 Amps Maximum 1.9 oz. Wt.

Motor, Battery: Manufacturer: AeroVironment, Inc. Type: Lithium Ion 13.5 Amp hour 22.2 V (nominal)

Fuel Not applicable NOTE: The Puma AE is powered by a Lithium Polymer rechargeable battery, AV Part Number 50318.

Engine (Propulsive Unit) Limits Maximum power output: 1.0 HP Maximum RPM: 10,000 RPM Maximum motor temperature: 180 °F NOTE: The motor temperature is managed by the Puma AE system and not displayed to the operator. Maximum motor, controller sub-assembly temperature: 194 °F (90 °C) Minimum voltage, motor battery during pre-flight engine run up after 3 secs at max throttle: 22.6 V Propeller and Propeller Limits (1) AeroVironment, Inc., Model 50330 FAA Propeller Type Certificate: None Propeller Type: 2-blade, hinged (folding), carbon fiber reinforced plastic, fixed pitch, tractor

Propeller Sub-Assembly: Manufacturer: Aeronaut Model: CAM 13 x 10 Diameter (Nominal): 13.3 in. Battery Command & Control Puma AE Air Vehicle Battery AV PN 50318 powers the motor, and battery command and control

Airspeed Limits VNE (Never Exceed Speed) 58 knot (30 m/s) VNO (Maximum Cruising Structural Speed) 41 knots (21 m/s) VA (Maneuvering Speed) 41 knots (21 m/s) Landing Speed: The landing configuration can be engaged (autoland) at any speed.

Center of Gravity (C.G.) Range 12.75 - 13.75 inches aft of datum

Empty Weight C.G. Range 12.75 - 13.75 inches aft of datum

Datum Front of motor case

Mean Aerodynamic Chord (MAC) 10.23 inches (259.8 mm) long with leading edge 9.11 inches (231.4 mm) from datum

Leveling Means Not Applicable

Maximum Weights Ramp 13.4 lbs. Takeoff 13.4 lbs. Landing 13.4 lbs.

Empty Weight 13.4 lbs.

Frequencies M1 (OCONUS) 1625-1725 MHz, M2 (CONUS) 1755 -1850 MHz Notes: FCC license is required to utilize the above frequencies; Uplink, downlink, and video are on the same frequency

Computer Software Motor Controller Interface Board Software: PN 58841 Revision A, Software Version 1.0.6 Avionics CPU, C-Code, Puma AE DDL PN 64321 Revision A, Software Version 52.02.27 Minimum Crew (1) The Puma AE system can be operated by a single operator. Number of Seats (0) Not Applicable

Fuel Capacity Not Applicable

Oil Capacity Not Applicable

Max. Operating Altitude 2000 ft. AGL (610 M)

Control Surface Movements

Nominal Endurance 120 minutes above 32 °F (0 °C) 60 minutes below 32 °F (0 °C)

Flight Limitations 1. Day Visual Flight Rules (VFR) in visual meteorological conditions (VMC) 2. Flight through visible moisture: PROHIBITED 3. Flight operations in icing conditions at assigned operational altitudes: PROHIBITED 4. Ambient Outside Air Temperature (OAT) a. Maximum OAT: 120°F/49°C b. Minimum OAT at Altitude: -20°F/-29°C 5. Wind. See Note 5. 6. Flight Operations. See Note 4. 7. For this operation only one Puma AE can be airborne at any given time. 8. Over water operation: PERMITTED 9. Over land operation: PROHIBITED 10. Only for operation in the designated Arctic Area as defined by the FAA Modernization and Reform Act of 2012, Section 332(d)(1). 11. An authorization for the specific location of operation issued by the Administrator is required and must be available at the control station. AFM number 72373_10X, dated July 16, 2013 or later FAA approved revision, and certificate of airworthiness (C of A) must be available at the control station (reference FAA Memorandum, "Certification of Unmanned Aircraft", from AAL-7 to ANM-100L, dated June 19, 2013). Additionally, any certificates of authorizations or waivers must be available at the control station. 12. Operation with inoperative instruments and equipment: PROHIBITED

Serial No. Approved 1723, 1726, 1728

Elevator Up 50° Down 24° Flaps Rudder N/A Left 45° N/A Right 45°

Certification Basis Restricted Category Only 14 CFR part 21.25(a) (2) for the special purpose of aerial survey, 14 CFR part 36, amendment 29, Appendix G

NOTES:

NOTE 1 Weight and Balance data are not applicable to the Puma AE. The aircraft operates in one configuration. The total aircraft weight with payload and standard equipment is 13.4 lb. NOTE 2 Placards Required: None

NOTE 3 This UAS must be maintained in accordance with AV Puma AE Maintenance Operation Manual, 72407_10X, or later FAA accepted revision

NOTE 4 UAS shall be operated under 14 CFR part 91, operating requirements, as mitigated. Operations shall be conducted in accordance with a waiver of flight regulations applicable to the operation, including but not limited to 14 CFR § 91.113, issued by the Administrator and specific to the intended operation, including geographical limitations.

NOTE 5 Wind Limitations: 25 knots NOTE 6 Personnel Keep Out Zones. Typical exclusion zones apply for Launch and Recovery as described in AFM and UAS Operations Manual.

NOTE 7 This Type Certificate Data Sheet (TCDS) is the principal document for Puma AE Operation. For any operational discrepancies among the TCDS, AFM 72373_10X and AV's Puma AE Operators Manual, 62869_A, etc., this TCDS takes precedence.

NOTE 8 Restricted category aircraft may not be operated in a foreign country without the express approval of that country.

NOTE 9 This aircraft has not been shown to meet the requirements of the applicable comprehensive and detailed airworthiness code as provided by Annex 8 of the Convention on International Civil Aviation. This aircraft meets14 CFR § 21.25(a)(2).

NOTE 11 Operations shall be conducted by properly certificated airmen who have completed training, checking, currency, and recency of experience requirements as approved by the Administrator.

NOTE 12 The Flight Standards Board (FSB) report is available on request. Contact the Long Beach AEG (LGB-AEG-NM17).

Appendix 7 200 Commercial Applications for Drones

Precision Agriculture

Plant Fertility Assess

Plant Water Content

Plant Disease Detect

Weed Mapping

Invasive Plants

Insect Attack Warning

Vegetation Identification

Selective Harvesting

Canopy Management

Herd Tracking

Telecommunications

High Altitude Imagery

Maritime Surveillance

Media

Traffic Monitoring

Disaster Relief

Real Estate Photography

Meteorology

Hurricane Monitoring

Cryospheric Research

Bridge Inspection

Transmission Line Inspect

HAZMAT Inspection

Emergency Medical Supply

Aerial Surveying

Damage Assessment

Insurance Claim Appraisal

Shark Watch

Concert Security

Sports Video

Runway Inspection

Virtual Tours

Coffee Harvest

Cinematography

Virtual Tours

Hydrologic Modeling
Flood Risk
Pollution Monitor
Tidal Zone Mapping
Anti-Piracy
Rail Track Bed Inspect
Saltwater Infiltration
Illegal Ship Bilge Venting
Terrain Mapping
Hydrometric Mapping
Highway Design
FedEx Unmanned Cargo
Advertising
Pavement Roughness
Prevent Extinction
Meteorology
Entomology
Fisheries Management
Wildlife Inventory
Remote Aerial Survey

Geomorphic Model
Law Enforcement
Photogrammetry
Solar Panel Inspect
Algae Proliferation
Ocean Research
Landmark Inspection
Emergency Com
Sand Bank Shift
Traffic Accident Analysis
Parking Utilization
Instant Consumer Grat
Coastline Surveillance
Animal Rights Groups
Ant-Whaling
Flood Warning
Forestry Inspection
Species Conservation
Mineral Exploration
Forest Fire Surveillance

Forest Fire Mapping

Remote Aerial Mapping

Snow Pack Avalanche

Poaching Patrol

Public Safety

Golf Resort Market

Training

Stadium Events

Power Restoration

Fire Prevention

Fire Risk Assess

River Discharge

Maritime Mammals

Pipeline

Chimney Inspection

Motion Pictures

Aerial Photography

Transmission Line

Real Estate Marketing

Aerial Land Survey

Surveillance

Volcano Monitoring

Oil Spill Tracking

Ice Pack Monitoring

Insurance Claims

Firefighting

Search and Rescue

Meteorology

Inspect Pipelines

Newspaper Delivery

Wind Turbine Blade

Marine Sanctuary

Ship Collision

Train Crashes

Pipeline Inspection

Inspect HAZMAT

Inspect Landmarks

Manage Fisheries

Inspect Icebergs

Crop Dusting

Volcano Monitoring

Predict Hurricane

Predict Earthquake
Archaeology
Com Relay
Oil Rig Inspection
Crime Investigation
Predict Earthquakes
Forest Management
Aerosol Measurement
Maritime Surveillance
Geophysical Survey
Archaeology
Crop Dusting
Windmill Inspect
Nuclear Inspection
Urban Planning
Preserve Culture
Oil Discovery
Discover Faults
Traffic Flow Analysis
Prevent Graffiti
Sports Event

Wildlife Research
Food Delivery
Crop Dusting
Predict Earthquake
Landslide Prediction
Railroads
Cloud Measurement
Coastal Water Quality
Work Documentation
Wildlife Research
Com Relay
Monument Inspect
Dam Inspection
Runway Inspection
Gravel Pit Inventory
Petroglyph Preservation
Predict Landslides
Highway Design
Crash Reconstruct
Poaching Patrol
Mineral Exploration

Journalism

Sea Level Rise

Fishing Enforce

Oil Spill Detect/Monitor

Tree Illness Monitor

Nuclear Radiation Meas

Training

VIP Security

News Coverage

Biological Agent Detect

Port Security

Forensic Photography

Entomology

Medical Supply Delivery

Atmospheric Profile

Coastal Patrol

Highway Roughness

Coastal Change

Seismology

Bank Erosion

Internal Tornado Measure

Volcano Ash Measurement

Construction Document

Meteorology

Roof Inspection

Wind Turbine Brake

Forest Health

Criminal Car Tracking

Crowd Control

Traffic Accident

Topographic Maps

HAZMAT

Event Security

Construction

Air Sampling

Avalanche Monitor

Customs & Border

Cloud Properties

Ozone Chemistry

Hurricane Genesis

Boston Marathon

Archeology

Climate Monitoring

Forest Regeneration

Tree Growth

Appendix 8 Unmanned Aircraft Professional Association

This organization is for anyone that desires to start a drone/unmanned aircraft (UA) business. A one stop shop for your business startup at www.ua-pa.com

UAPA provides help to those that would like to gain knowledge about the commercial applications for unmanned aircraft (UA) and have or would like to start a Unmanned Aircraft (UA)/Drone business in operations, services, apps development and manufacturing. With your involvement and strong support we can address the critical issues associated with commercial unmanned aircraft (UA)/Drone applications. Many people already operate successful businesses in this emerging area. Anyone, anywhere in the world can join. Get in on the ground floor of this profitable emerging industry. Your membership gives you full access to all of our benefits and resources including:

- **Legal Representation**
- **Business Insurance**
- **Training**
- **FREE Business Start-Up Counseling Session ($100 value)**

- **Discount on Annual Members Meeting**
- **Member Certificate**
- **Membership Card**
- **Lapel Pin**

Legal Representation: Obtain legal services from our partner law firms that specialize in UA/Drone operations, services and manufacturing.

Insurance: Select from a list of insurance services from our partner insurance firms that specialize in UA/Drone operations, services, apps development and manufacturing.

Training: Learn the theory and application from our academic courses or obtain UA/Drone Pilot training. Purchase a course on DVD, participate in an online course or visit one of our UA pilot training schools.

Business Start-Up Counseling: A member can select a UA/drone business application and ask for a business case analysis. UAPA will provide market analysis and recommendations.

Unmanned Aircraft Professional Association (UAPA), International provides numerous services for start-up commercial UA/Drone companies. We are hoping to recruit over 500,000 members which will make UAPA the largest aviation organization in the world.

The pledges you make will provide you with training courses to teach you the basics of UA/Drones so you can either obtain employment or start your own UA/Drone business. This can be done reasonably today because costs continue to decrease. Joining UAPA will provide you with the business start-up advice, legal representation and insurance to make your business a success.

Unmanned Aircraft Professional Association (UAPA), International is a one stop shop for UA/Drone business owners. This is an International Association open to everyone worldwide. To join go to www.ua-pa.com

Appendix 9 Small Drone Sensor Manufacturing Companies

Infrared	
FLIR Tau 2 and Quark	flir.com
IAI MicroPop	iai.co.il
IRCameras	ircameras.com
DST Control	dst.se
Mutispectral	
Tetracam ADC Micro and Mini MCA	tetracam.com
Quest Innovations	quest-innovations.com
FluxData	fluxdata.com
BAE Systems	baesystems.com
Pixelteq	pixelknowledge.com
Hyperspectral	
Headwall Photonics	headwallphotonics.com
Hyspex	hyspex.no
Opto Knowledge	optoknowledge.com
LIDARoptoknowledge.com	
Advanced Scientific Concepts	advancedscientificconcepts.com
Velodyne	velodyne.com
Autonomous Stuff	autonomoustuff.com
Radar	
IMSAR	imsar.com
Ultrasound	
Maxbotix	maxbotix.com
Custom Sensors	
A2e Technologies	a2etechnologies.com

Appendix 10 Small Drone Manufacturing Companies

Company	Website
Accurate Automation	accurate-automation.com
Acuity Technologies	acuitytx.com
Adcom Systems	adcom-systems.com
Aeromapper	aeromao.com
AESIR	aeromao.com
Adaptive Flight Inc	adaptiveflight.com
Aerosight	civilianuav.com
Aerovision	aerovision-uas.com
AeroVironment	avinc.com
Aeryon Labs SkyRanger	aeryon.com
AiDrones	aidrones.de
Airborne Technologies Inc	airbornetech.com/
Airbotix	airbotix.com
Air Robotics	airrobotix.com
Airship Manufacturing	airshipmanufacturing.com
Altavian	altavin.com
American Aerspace Airborne Systems	american-aerospace.net
American Dynamics	americandynamics.net
American Unmanned Systems	americanunmannedsystems.com
Arcturus UAV	arcturus-uav.com
Aurora Flight Sciences Skate	aurora.aero
Applewhite Aero	applewhiteaero.com
Bird Aerospace Bird's Eye	birduas.com
Brock Technologoes	brocktechnologies.com
Bosh Technologies	boshtech.com
Cat UAV	catuav.com
Crescent Unmanned Systems	crescentunmanned.com
Cropcam	cropcam.com
CyberAero	cyberaero.se
CybAero	cybaero.com
Delair-Tech	delair-tech.com

DJI Innovations	www.dji-innovations.com
Dragonfly Innovations	www.dragonfly.com
Dragonfly Pictures	dragonflypictures.com
Droidworx	www.droidworx.co.nz
Drone America	droneamerica.com
Ecilop Flying Camera	www.ecilop.com
Falcon UAV	falcon-uav.com
Fan Wing	fanwing.com
Fly-N-Sense	fly-n-sense.com
Flight Evolution	flighteevolution.com
Frontline Aerospace	frontlineaerospace.com
Hawkeye UAV	hawkeyeuac.com
Hoverfly Technologies	www.hoverflytech.com
Innovative Automation Technologies	iat-llc.com
Insitu Scan Eagle	insitu,com
Israel Aerospace Industries	www.iai.co.il
Marcus UAV Systems	marcusuav.com
Maryland Aerospace	imicro.biz
MicroKopter	www.mikrokopter.com
Microdrones	microdronescom
MLB Company	spyplanes.com
Northrop Grumman	northropgrumman.com
Novadem	novadem.com
Parrot	ardrone2.parrot.com
Photohigher	photohigher.co.nz
Piasecki Aircraft	piasecki.com
Precision Hawk	Precision Hawk
Prioria	prioria.com
Pulse Aerospace	pulseaero.com
Quest UAV	questuav.com
Rotomotion	rotomation.com
Scion UAS	scionuas.com
SenseFly	sensefly.com/home.html

Shadowair	shadowair.com
Silent Falcon UAS Technologies	www.silentfalconuas.com
Tean Black Sheep	team-blacksheep.com/shop
Trigger Composites	kompozyty.trigger.pl/EN/
Trimble	www.uas.trimble.com/
TOR Robotics	torbotics.com
Turbo Ace	www.turboace.com
UAS Europe	uas.europe.se
UAS Solutions	uas-soluton.com
UAV Factory	uavfactory.com
Unmanned Systems Group	unmannedgroup.com
Vanguard Defense Industries	vanguarddefense.com

Appendix 11 Small Done Component Companies

Antennas	
Dayton-Granger	dayton-granger.com
Antcom	antcom.com
Autopilots	
Airware	airware.com
MicroPilot	micropilot.com
Piccolo	cloudcaptech.com
Procerus/Kestrel	procerusuav.com
Batteries	
Hobby Lobby	hobbylobby.com
MaxAmps.com	maxamps.com
Engines	
UAV Engines	uavenginesltd.co.uk
Hirth Motors	hirth-morten,de
Geospatial Products	
Isis Geomatics	isisgeo.com
Overwatch	overwatch.com
Orbit Geospatial	orbitgis.com
Ground Control Station Software	
Instrument Control Sweeden	instrumentcontrol.se
Micropilot	micropilot.com
Kutta Technologies	kuttatech.com
Navigation	
Geodetics	geodetics.com
Sagetech	sagetechcorp.com
Trimble	trimble.com
Power Supplies	
Abbott Technologies	abbot-tech.com
Advanced Conversion Technology	actpower.com
Transmitters/Receivers	
AMP Advanced Microwave Products	advmw.com
Teletromics Technology Corporation	ttcdas.com

Appendix 12 Small Drone Services Companies

Data Processing	
Aegis Technologies	aegistg.com
2D3 Sensing	2d3.com
Precision Hawk	precisionhawk.com
Robota	robota.us
Altavian	altavia.com
Ersi	esri.com
Pix4D	pix4d.com
MosaicMill	support@mosaicmill.com
Integration	
Aero Surveillance	aerosurvellance.com
Leasing	
Aurora Flight Sciences Centaur	aurora.com
Advanced Unmanned	advancedunmanned.com
LIDAR	
Phonix Aerial Systems	phoenix-aerial.com
Operations	
ISR Group	isrgroup.com
Remotely Piloted Solutions	rps-uas.com
Simulation	
CAE	cae.com
VT MAK	mak.com

Appendix 13 UAV Insurance Providers

Costello Insurance	www.aviationi.com
Transport Risk Management	transportrisk.com
Aerial Pak	www.aerialpak.com
John Heath Insurance Brokers LLP	www.johnheath.com
R J Kiln & Co Ltd	www.kilngroup.com
Overwatch	www.riskoverwatch.com
Catlin	www.catlin.com
Visionair	www.visionair-uk.com
ABSaviation	www.absaviation.com
Bill Owen Insurance	www.billowen.com.au
Avro	www.averoins.com
Avpac	www.avpac.com

Appenix 14 Unmanned Vehicle Univerity Press

Unmanned Vehicle University Press is a Puplications Division of Unmaned Vehicle University that publishes books and courses on Unmanned Systems. To encourage more publication, UVU Press is offering free publication for the first 50 authors. UVU will review your book and if approved, provide the following services for free:

Create a print ready proof for your approval
Create a cover
Hardcover or softcover
B&W or color
Choice of book size
Includes 13 digit ISBN number
Includes Library of Congress Number
Complete the publication process
Available from Amazon, Barnes and Noble, Baker and Taylor and Ingram
Advertise in 100 countries
Advertise to over 30,000 wholesalers retailers and booksellers
Set your own price and receive royalties on every book.

How it works

1. Write the book
2. Save to pdf format
3. Send as email attachment to submit@uvupress.com
4. See the contents of the email at uvupres.com
5. Begin review process, optional services and cover design
6. Proof and cover design mailed to author for approval
7. Book published on Amazon, B&N, Baker & Taylor
8. Royalties paid to author on quarterly basis

The hard part is writing the book, publishing it shouldn't be.

www.uvupress.com

CPSIA information can be obtained at www.ICGtesting.com
Printed in the USA
BVOW11*0526240314

348491BV00001B/1/P